SpringerBriefs in Space Development

Series Editor

Joseph N. Pelton, Jr.

Guest Editor

William H. Ailor

For further volumes:
http://www.springer.com/series/10058

Joseph N. Pelton

Space Debris and Other Threats from Outer Space

Guest Editor: William H. Ailor

Joseph N. Pelton
Executive Board
International Association for the
Advancement of Space Safety
Arlington
USA

ISSN 2191-8171   ISSN 2191-818X   (electronic)
ISBN 978-1-4614-6713-7   ISBN 978-1-4614-6714-4   (eBook)
DOI 10.1007/978-1-4614-6714-4
Springer New York Heidelberg Dordrecht London

Library of Congress Control Number: 2013933569

Printed on acid-free paper

Springer is part of Springer Science+Business Media (www.springer.com)

ISU (Society) Page

This Springer book is published in collaboration with the International Space University. At its central campus in Strasbourg, France, and at various locations around the world, the ISU provides graduate-level training to the future leaders of the global space community. The University offers a two-month Space Studies Program, a five-week Southern Hemisphere Program and a one-year Masters program related to space science, space engineering, systems engineering, space policy and law, business and management, and space and society.

These programs give international graduate students and young space professionals the opportunity to learn while solving complex problems in an intercultural environment. Since its founding in 1987, the International Space University has graduated more than 3,000 students from 100 countries, creating an international network of professionals and leaders. ISU faculty and lecturers from around the world have published hundreds of books and articles on space exploration, applications, science and development.

Contents

Chapter 1
Serious Threats from Outer Space

This must be the context of our thinking…the vast new dimen-
sions of our science and our discovery…and of the awful
majesty of outer space.

—Adlai Stevenson at Harvard University (1965)

What Space Threats Should Concern Us Most?

Have you ever wondered if you might be hit by a piece of space junk falling out of the sky? Well not to worry. The chances of being hit by a piece of orbital debris or a meteorite are fewer than your getting the world's rarest disease or being killed by a falling coconut if you only visited a tropical paradise for only 1 h in your entire lifetime. In over a half century of space activities there is perhaps one instance of a cow being killed in Africa and that was many decades ago. More recently in 1997 a woman was reportedly brushed on the shoulder by a lightweight fragment of space debris falling from the skies.

But there is still a very good reason for you to read this book. In fact, there are serious threats from space that are actually of mounting concern that you should know about and what actions could be taken to forestall these threats. The growing amount of space debris in the skies could make it difficult in the future to access space for many crucial applications, such as communications, navigation, remote sensing, weather forecasting, military surveillance, nuclear monitoring, or even space exploration. A surprising amount of the world scientific, economic and military activities are now based on spacecraft operations. Solar radiation that penetrates tthe ozone layer can and indeed does cause skin cancer, and currently occurring changes to the ozone layer are elevating this concern. The most intensive cosmic radiation from gamma rays if unchecked and sustained can also trigger harmful and even bizarre mutations of our genes so as to prevent healthy and normal reproduction.

Solar flares or coronal mass ejections could kill astronauts or wipe out our electrical grids in a powerful and instantaneous way, as happened as recently as

J. N. Pelton, *Space Debris and Other Threats from Outer Space*,
SpringerBriefs in Space Development, DOI: 10.1007/978-1-4614-6714-4_1,
© Joseph N. Pelton 2013

March 1989. Although very unlikely, a massive and potentially harmful near-Earth asteroid could destroy much of life on Earth, as was the case with the so-called K-T event (i.e., the Cretaceous Tertiary Mass Extinction Event) that fortunately occurred some 65 million years ago. This single calamitous cosmic occurrence caused between 65 and 70 % of all species on Earth to be killed—including the dinosaurs—in what is known simply as a mass extinction event [1]. On the scale of bad things to happen, this would be very, very bad indeed.

However, let's start at the beginning to recap what has happened since the Space Age began. Let's quickly review why today we know much more about cosmic threats than we ever did before, and why in learning about space through sending probes aloft we have managed to create some serious new problems of our own making.

At the Beginning of the Space Age

The age of spaceflight began well over five decades ago on October 4, 1957, with the launch of *Sputnik*. When this first spacecraft was launched into Earth orbit, it was hailed as a major advance in human scientific and engineering achievement [2]. The stark realities of the Cold War between the Soviet Union and the United States, however, also painted this first space launch in a vivid military context as well. This Soviet coup in space jolted a vibrant American space program into action. In just a few years there were a number of spacecraft and missile launches occurring in both the United States and the U.S.S.R. [3].

Back in 1957 little thought was given to what might be the risks associated with *too* many spacecraft launches. Over a half century later, however, the accumulation of human-built space debris in orbit is now a quite real problem. "Space junk" is now increasingly seen as a creditable threat to humanity's longer term ability to access and utilize space. In literally dozens of ways humanity is dependent on satellites to communicate, to navigate, to track killer storms, and to provide an effective military defense capability. "Space junk" every day and in every way is becoming a true threat.

If we could effectively stop the creation of all new space debris, we would still not have solved the problem. In fact, the accumulation of debris, just due to collisions from existing space junk, 50 years hence would still be significantly worse. But in fact we are still launching more and more satellites, and space debris continues to mount.

Visionary Ideas are Easy to Often Easy to Dismiss

About a quarter of century ago the possibility that space debris might constitute a tangible threat to our longer term space programs was raised by space scientists and especially by Donald Kessler. Unfortunately at the time this was largely treated as simply a laughable idea. No one is laughing now. Figure 2.1, later in the book, shows rather graphically how this problem seemed to sneak up on us over the past few decades. Human skepticism often serves us well, but sometimes it smothers the most important new ideas. In the area of space this has often been the case.

Robert Goddard, the father of modern rocketry, and other innovators have taught us that there is often a thin line between longer-term vision and what are generally considered as outlandish or quixotic ideas.

In 1919 the Smithsonian Institute published a report by Robert Goddard outlining his plans to launch liquid-fueled rockets. In this treatise he indicated how such rockets could eventually reach the Moon. In 1920 the *New York Times,* with more arrogance than scientific knowledge, responded by running a derisive editorial to call Robert Goddard "The Moon Man" for his audacious claim that one day rockets would carry human adventurers to the lunar surface. Goddard persevered and successfully launched his first liquid-fueled rocked on March 26, 1921. Goddard famously said: "Every vision is a joke until the first man accomplishes it. Once realized, it becomes commonplace." But it was not until a day after the Apollo Moon landing in 1969 that the *New York Times* ran a correction and an apology for its errors in the 1920 editorial—albeit some 49 years late [4].

Today space debris is no longer a "laughable idea" More and more people will have the opportunity to fly into space on governmental and commercial spacecraft in the twenty-first century. Up until the end of 2012 only about 500 people have flown into space. As twenty-first century commercial space industries mature, we will actually see more and more "citizen astronauts" flying on sub-orbital flights or even going into orbit. Unfortunately, for all future astronauts, whether government or private spacefarers, the risks to them from space debris will mount as they ride on rockets or live aboard space stations. For most people who will never venture into space, there are still areas of concern. The space debris actually does come down and sometimes at unfortunate times and places.

The Mounting Problems of Space Debris

Right now the biggest risk is that vital communications satellites or other key spacecraft can be destroyed by space debris traveling at speeds in excess of Mach 20. The ability of space debris to knock out spacecraft or injure or kill astronauts in space must now be taken seriously. There is also a concern that falling debris could cause property damage or even kill, but this probability fortunately is very, very small. The point is that now is the time to address all of these concerns.

How was this problem created? Over a period of time more and more space launches occurred. With these launches various types of debris began to accumulate. There are now explosive bolts, exploded fuel tanks, paint chips, upper stage rockets, rocket fairings that covered satellites that were being deployed in higher orbits, defunct satellites, and finally—in the last few years—debris from colliding satellites and even debris from a defunct satellite deliberately being hit by a ground-based missile.

At first there was only a minor amount of debris. But over the decades the debris accumulated. In time scientists began to understand that all this debris was beginning to pose a serious risk and a spreading "pile of space rubble" was accumulating—particularly in certain orbits. This now huge amount of space junk has now begun to threaten human longer-term access to space. Just as we now worry about the "sustainability" of life on Earth due to greenhouse gases and over population, we are worried about the "sustainability" of access to space due to space debris.

An enormous quantity of human-made debris is swirling around Earth, particularly in low Earth orbit (LEO). Scientists have determined that there are literally millions of debris elements in Earth various orbits—primarily LEO, but also Medium Earth Orbit (MEO) and geosynchronous earth orbit (GEO)—all of which have begun to fill up with space junk.

Many of these elements—literally millions of them—are of microscopic size and involve things such as chips of paint. It is currently estimated, however, that there are between 500,000 to 750,000 objects in orbit that are on the order of 1 cm in size. The first reaction of most people is something like, whew, those are really little guys that surely cannot do much harm.

Figure 1.1 shows a 1-cm puncture in the high gain antenna on the Hubble Space Telescope. A chip of paint traveling at 17,000 mph or over 28,000 kmph can put a serious crack in the window of a space shuttle or rupture an astronaut's spacesuit. An element as large as 1 cm can do substantial harm, and

Fig. 1.1 Puncture in Hubble Space Telescope array cause by space debris (Image courtesy of NASA)

something as large as 10 cm (4 inches in size) could potentially destroy a communications or remote-sensing satellite or some other valuable space resource. Shielding or armor on satellites against debris is really effective only up to about 1 cm.

The Challenge of Tracking Space Debris

In 1980 there were just fewer than 5,400 sizeable objects (i.e., greater than 10 cm in size spinning around in low-Earth orbit) that were being actively tracked. By 2010 the number of large space debris objects had increased to 15,639. Today there are some 22,000 objects that are 10 cm or larger being tracked by the U. S. Air Force Space Surveillance Systems (AFSSS). This is a combination of ground-based plus several satellite-based tracking systems.

The current Very High Frequency (VHF) radar is being upgraded by a new "space fence" radar system operating in the S-band that will provide much greater resolution. (See Fig. 1.2) The Air Force Space Surveillance System which was first implemented in 1961, in part as a missile tracking system that is now aging. Thus the Air Force has now contracted for a debris tracking system that is to be fully implemented by 2017. Tests carried out in March 2012 confirmed the new tracking

Fig. 1.2 U. S. Air Force satellite used for space debris tracking (Graphic courtesy of the U. S. Air Force)

capabilities and the effectiveness of the overall design of this so-called Space
Fence by accurately tracking space debris elements. The details of this system will
be discussed in greater detail in Chap. 2 [5].

Of the 22,000 objects being tracked by the current AFSSS about 1,000 objects
represent functional satellites, but the rest are "defunct" satellites or other forms of
space junk.

The largest pieces of debris are most important to track for at least two reasons.
First, these bigger objects can literally destroy the International Space Station (ISS)
or other billion-dollar space facilities because of their huge kinetic energy, equivalent to large bombs. Secondly the collision of large space objects—regardless of
their operational status—can create perhaps many thousands of major new debris
elements. Big space objects colliding with each other is the number one problem
we must seek to avoid, although it is imperative to find ways to reduce the formation of any type of new debris as well as a way to remove orbital debris from orbit
in a systematic way regardless of size.

Over 6,300 Tons of Debris in Earth Orbit

The build-up various sized debris elements over the past two decades has now
become alarming. The following chart from NASA explains the size of the various types of debris and their relative distribution. Fortunately most of the hundreds
of millions of debris elements now in Earth orbit represent microscopic elements
such as chips of paint. These microscopic elements are just the size of a grain of
salt, but when traveling at speeds of perhaps 28,000 kmph (or about 17,500 mph),
still pack quite a wallop, a wallop sufficient to penetrate the spacesuit of an astronaut or perhaps pit or even penetrate a window on a space vehicle. (See Fig. 1.3).

The nature of this problem, i.e., big objects colliding, has been vividly demonstrated within the past decade. First there was the collision of the operational
Iridium 33 mobile communications satellite and the defunct Russian Cosmos 2251
weather satellite. Then there was the occasion when a Chinese anti-missile deliberately hit a defunct Chinese weather satellite. In both events on the order of 3,000
new tracked debris elements were created and resulted in a new level of threat
to the International Space Station. In short these two events led to an impulse
increase in "trackable" space debris objects by some 6,000. The Fig. 1.4 creates
a representation of the debris created by the missile hit on the Feng-yun weather
satellite and how this new swarm of debris relates to the orbit of the International
Space Station as represented by the "white orbit" in the illustration.

The greatest concern with regard to space debris is the so-called "Kessler
Syndrome". This is a condition whereby colliding space junk creates a deadly ongoing avalanche of more and more debris elements. Space scientist Donald Kessler in
1978 wrote a paper that warned that this type of problem could actually occur [6].

His paper explained how an ongoing series of collisions of space debris could
lead to a cascade effect whereby the problem would become worse and worse

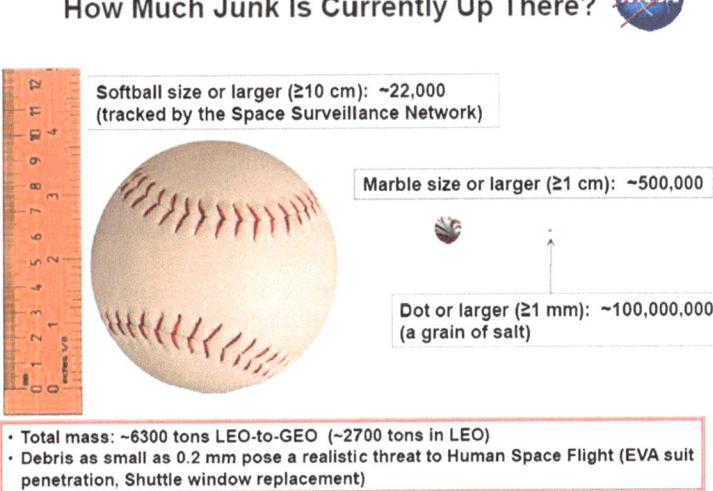

Fig. 1.3 Breakdown of the 6,300 tons of mass in Earth orbit (Graphic courtesy of NASA)

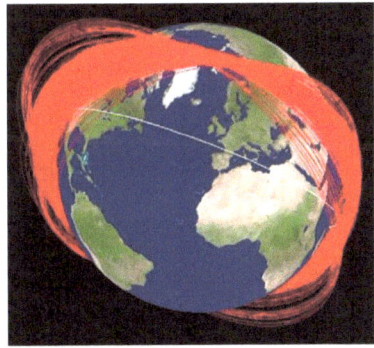

Fig. 1.4 Artist representation of orbital debris created by the destruction by missile of the Feng-yun weather satellite (Graphic courtesy of NASA Space Debris Program Office)

once a "tipping-point" had been reached. Kessler's warning—now known as the Kessler Syndrome—explained that once this tipping point was reached the problem would grow out of control. His early predictions of this effect, however, were not taken too seriously.

The truth is that important forecasts about space, from those of Sir Arthur Clarke concerning global satellite communications to those of Robert Goddard about lunar vehicles with human crews, were ridiculed or ignored when first made. Now, as the space debris problem has grown just as Kessler forecast, this problem is widely acknowledged around the world [7]. In fact, a report by the U. S. National Research Council in September 2011 concluded that the problem was "worse than had been early thought" [8].

Currently, the only mechanism for removal of debris is orbital decay through atmospheric drag and Earth's gravitational attraction, which ultimately leads to re-entry. Unfortunately, such gravitational removal of debris only works effectively for low-Earth orbits. For satellites in medium Earth orbit above the Van Allen Belts, it takes hundreds to thousands of years for objects to re-enter Earth's atmosphere. For geosynchronous orbits, that are essentially one-tenth of the way to the Moon and where the pull of gravity is only 1/50th of that at Earth's surface, the gravitational decay process for debris elements is essentially negligible. For a geo satellite to come down would literally take many millions of years. Consequently, there is currently no effective removal mechanism for MEO or GEO debris elements unless there were to be active rockets designed for controlled de-orbit.

As noted in Fig. 1.3 the build-up of debris elements in Leo orbit particularly in the polar area has now reached the incredibly high number of 2,700 tons, which far exceeds the gravitation degradation of a few tens of tons a years. Figure 1.5 shows that Leo polar orbits in particular are now extremely congested. This figure shows in some detail the debris that is being tracked in LEO by the U. S. tracking system.

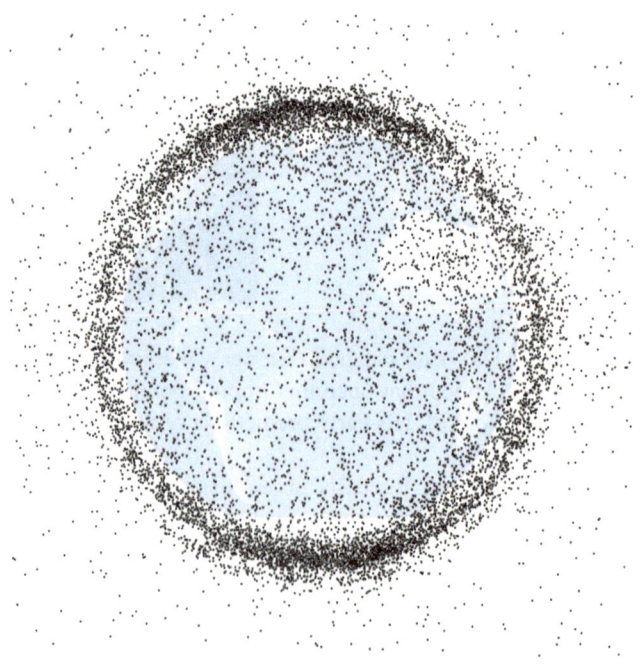

Fig. 1.5 Debris elements in LEO pictured over South America and Africa (Image courtesy of NASA Space Debris Projects Office)

Debris is Increasing Faster than its Decay

Historically, the creation rate of debris has outpaced the removal rate by a fairly wide margin. This is leading to a net growth in the debris population in LEO at an average rate of approximately 5 % per year. Although this may not sound like much, it means the amount of debris in orbit is now very substantial and based on past experience will likely continue to grow.

Although the low-Earth debris orbits in effect "spread out" as they orbit Earth they come much closer together over the North and South Polar Regions and thus serve to increase their chances of collision by a considerable degree.

To put the problem into a more accurate "visual perspective," it important to note that because the scale used in Fig. 1.4 is actually something like 10 million to one, the risk of collision is indeed much less than it would appear. It's like when one looks in a rear view mirror and it says that "vehicles may be closer than they appear". Recognize that the same is true here. The debris elements shown above, in fact, are 10 million times further apart and that the volumetric space of Earth and space within which the orbiting debris is depicted here is 10^{21} times (or 1,000,000,000,000,000,000,000 larger).

A major contributor to the current debris population has been fragment generation via explosions of fuel tanks and more recently by collisions. It is hoped that future explosions can be minimized by venting of fuel prior to the operational end of life of satellites—as recommended by the current mitigation procedures. It may take a few decades for the practice to become implemented widely enough to reduce the explosion rate, which currently stands at about four per year.

Several environment projection studies conducted in recent years indicate that, with various assumed future launch rates, the debris populations at some altitudes in LEO will become perhaps completely compromised. In these projections collisions could take over as the dominant debris generation mechanism, and the debris generated will feed back into the space environment and induce more collisions—in short, an in-orbit cascade that creates more and more debris.

According to studies conducted by J-C Liou and N. L. Johnson, the most active and endangered LEO region is between the altitudes of 900 and 1,000 km, and, even without any new launches, this region is highly unstable. It is projected that the debris population (i.e., objects 10 cm and larger) in this "red zone" will approximately triple in the next 200 years, leading to an increase in collision probability among objects in this region by a factor of ten [9]. In reality, the future debris environment is likely to be worse than as suggested by Liou and Johnson, as satellites continue to be launched into space. In late June 2012 this author was at the Kennedy Space Center where the Delta IV Heavy vehicle launched a surveillance satellite into orbit and this satellite with on-board positioning fuel alone weighed some 30 tons.

The Liou and Johnson paper concludes that to better limit the growth of future debris populations, active debris removal (ADR) from space needs to be considered. The various technical and operational options that are being considered for such removal are discussed later in this book.

The problem of tracking debris, of course, becomes more difficult as one moves further away from Earth in higher orbits. In the geosynchronous geostationary orbit, for instance, the minimum size that can be tracked is 30 cm in contrast to about 10 cm in LEO. Among the tracked pieces of debris, there are about 200 satellites abandoned in geostationary geosynchronous orbits occupying or drifting through valuable orbital positions and posing a collision hazard for functional spacecraft. Fortunately, accurate tracking systems, charting of possible conjunctions that could result in high velocity collisions, and active collision avoidance maneuvers minimize these risks. The Satellite Data Association (SDA) that will be discussed later now plays a key role in this activity.

As noted earlier the survival time of the debris in orbit continues to changes with the higher orbits. Objects in 1,000-km orbits can exist for hundreds of years. At 1,500 km, the lifetime can go up to thousands of years. Objects in geosynchronous or super synchronous orbits can survive for millions of years.

And there are other realistic space threats that also need to be taken seriously. Although space debris has now become a top issue that must be dealt with in order to sustain useful access to space, this is just one of the "threats" that must be addressed. The harsh environment of space puts satellites, space stations, and even rocket launchers at risk. These risks include, micro-meteorites, solar flares, coronal mass ejections (CMEs), and cosmic radiation. These natural hazards can disable or totally destroy functioning satellites and spacecraft as proven by past events. These events are less under our control than space debris, but shielding and other protective actions can help protect against these types of hazards as well. Currently these natural threats pose a higher risk level than space debris, but over time space junk, unless aggressively attacked by Active Debris Removal (ADR), will become a higher level threat.

Potentially Hazardous Asteroids and Mass Extinctions

What is often not mentioned is that these natural debris and natural phenomena could actually pose threats even to people on the ground. Solar flares, coronal mass ejections, cosmic radiation, meteorites, asteroids and comets, and yes, even space debris can pose risks to people on the ground. These risks to people right here on Earth's surface will be addressed in later chapters of this book. Most of these risks would involve only a limited number of people.

But there is one type of natural hazard that not only threaten astronauts and spacecraft but could indeed threaten life on Earth in a big way. This threat is known as Potentially Hazardous Asteroids (PHAs), and this is in no way just a "theoretical" risk. Actually this is something to be taken quite seriously. It is believed that an asteroid, rich in the poisonous substance iridium and perhaps 10 kilometers in diameter, plunged into Earth some 65 million years. When it impacted Earth it created a huge cloud around Earth that blocked out the Sun for several years. As a result the dinosaurs and well over a third of all life-forms on the planet died off. Another asteroid or large comet could do equal damage to humans and other life forms if it were to hit Earth in future years.

Even a smaller asteroid such as Apophis, which is about 300 m in diameter, if it were to hit in an ocean near a large city could bring death to tens of millions of people, and if it were to hit in, say, the United States it could possibly wipe out an entire state. Fortunately Apophis is scheduled to fly by in 2029 and 2036 and then be on its way [10]. This very real subject of "killer asteroids and comets" and what we are doing to be ready for them, will also be addressed in later chapters.

In short after addressing the problem of growing amounts of space junk the discussion will turn to various types of natural threats in and from space and even potential threats to people here on the planet's surface. In all cases the discussion will go beyond identifying risks to explore protective actions. It is not enough to just explain that there are threats. There are indeed a number of actions being taken to protect the billions of dollars in space assets from both space debris and natural space hazards. In fact, military satellites deployed in strategic regions are even hardened against nuclear weapon explosions, electronic magnetic pulses (EMPs) and cosmic radiation. As new techniques are developed to protect space assets and extend space situational awareness, these solutions can presumably be applied to help protect people here on Earth as well.

Purpose of the Book

The purpose of this book is to provide a good overall understanding of the nature of the various space threats and what techniques, new technologies and strategies can be developed to cope with these various hazards.

In addition there are programs operated by space agencies and research centers around the world related to protection of humanity against natural threats from space. These include:

- Operation of sophisticated systems to monitor solar activities such as solar events that can generate hazardous "space weather" (i.e., solar energetic particles—SEPs—and coronal mass ejections—CMEs—as well as cosmic radiation from the Sun and beyond).
- Intensive use of space telescopes and sensors and ground observatories to the orbits of asteroids and comets.
- R & D activities to develop systems to cope with potential "killer asteroids".

Despite all of these activities, there is evidence that what is being done may well not be enough.

Structure and Highlights of the Book

The structure of this book is to first introduce the nature of the problem of space threats and to note that the methodological approach to the subject is completely multi-disciplinary. Thus the technical, operational, economic and financial, and legal aspects of the problems related to space threats will be addressed along with possible solutions in each of these areas. In some cases an interdisciplinary approach is used simply because the solution may require new technology, new international legal regulations and financial incentives or penalties if corrective action is not taken.

Four chapters of the book provide a good deal more information about the various problems associated with space debris. These chapters address the technical, operational, institutional and even financial and regulatory arrangements associated with attempts to address and mitigate this growing and increasingly very real problem. The remaining chapters of the book address the very real threats that exist in space that come from natural space phenomena, including coronal mass ejections, solar flares, solar and cosmic radiation, and finally potentially hazardous near-Earth objects (NEOs), including comets and asteroids. Here is a quick recap with some key highlights.

Chapter 2 will address in depth why the threat of space debris and the Kessler Syndrome is increasing. This chapter explains that even if there were to be no new space debris created from new launches that the problem would still keep increasing for decades to come just from the space debris that is already out there. Chapter 2 also seeks to provide a general understanding as to how and why the problem of space debris will increase over time. This analysis notes we need to develop not only new technical and operational solutions, but also new regulatory, institutional and financial mechanisms and procedures as well.

Chapter 3 addresses the nature of the space debris problem and possible solutions that actually vary fairly widely in terms of the various orbits. The biggest and most urgent problem involves LEO and Sun-synchronous polar orbiting satellites as the top priority. Despite the fact that there is a critical need to get large space junk out of LEO, we should not lose sight of the need to clean up all the space around Earth in all the orbits. In short, solutions and corrective actions for all types of orbits from LEO out to geosynchronous orbit must be eventually found and implemented.

Chapter 4 addresses the institutional and regulatory issues. In particular, this chapter presents the specific efforts of the Inter Agency Space Debris Coordinating Committee (IADC) and the United Nations Committee on the Peaceful Uses of Outer Space (COPUOS). These international bodies have been seeking for some time to address the problems of space debris and the longer-term sustainability of space. So far they have evolved to the point of "voluntary guidelines" to minimize orbital debris. But we need to go much further. In addition to these two key international bodies there are other organizations and activities that are helping to develop improved space situational awareness and to coordinate activities among space system operators to avoid possible collisions. Two examples of such organizations are the Space Data Association and the U.S. Air Force Space Command that provides the prime space tracking capability.

The longer-term sustainability of space currently starts with the development of improved tracking capabilities. But this is only the start of the process. There are a series of legal, regulatory and liability issues related to orbital debris and space operation concerns that applies to all current and future space faring nations.

The current international liability provisions related to spacecraft and orbital debris, unfortunately, do not help with efforts to remove orbital debris from orbit. In fact, the current international liability convention might well be considered a barrier to this process. Indeed that is the opinion of most space legal experts that have addressed this problem. Most recently the COPUOS in 2011 established a Working Group on the Longer Term Sustainability of Space that examines the various issues related to making sure that all nations have the future ability to use space in a productive and effective way. This working group is addressing all of the technical, operational, and legal matters that are involved.

Chapter 5 addresses space debris remediation processes and the current status of space technology and related ground systems that might be employed to undertake space debris removal. In general, none of the technologies are really mature. Even if these various methods could be brought to technical and operational maturity they do not currently constitute cost-effective means to accomplish the task. In short a great deal of future research is needed in these areas to develop effective, cost-efficient methods for orbital space debris mitigation and also to avoid anything that might seem to be employing the use of "space weapons". In fact, finding ways to accomplish space debris removal with technology that would not be considered as a space weapon is one of the key challenges to overcome.

We next move on from space debris related issues to the very real concerns of natural space hazards and the problems and issues related to the so-called phenomena known as "space weather," cosmic radiation and potential collision with asteroids or comets. Here we explore the fact that the 'natural threats' from space endanger both our spacecraft in orbit and actually can endanger us here on Earth as well.

These natural space phenomena certainly include hazards to spacecraft and space operations. Satellites must be designed to withstand the very real difficulty of long-term operation in the very harsh space environment, where in-orbit repair or refurbishment is generally not an option. But the hazards involve more than just designing satellites to withstand the rigors of space and thus we will explore why and how we need to protect modern electronic infrastructure from space hazards as well. Although space debris is a very real threat to the long-term sustainability of space-related activities, it is important to understand that there are a number of very real natural space threats as well.

The hazards addressed in the later chapters actually could represent a much larger threat to humanity than space debris—and by several orders of magnitude. But fortunately that is not the whole story. Although the threat levels are high, the chances of many of the most hazardous events actually occurring—as triggered by natural space phenomena—are quite small. One of the great challenges for space scientists today is how to deal with threats that are very large, but with their chance of occurring being quite small.

Fortunately the protective shield of the Van Allen Belts, the ozone layer, Earth's atmosphere and especially Earth's geo-magnetosphere provide us critical life-saving protection. There are, however, currently two emerging problems in terms of Earth's protective system against threats from space. One problem is that Earth's geo-magnetic field seems to be developing "cracks" that could let highly destructive radiation and ionic particles as well as poisonous gases through with deadly effect. This is a problem being studied by space probes with some urgency. The other problem is what to do if Earth's protective atmosphere begins to rise to unacceptably high temperatures as the result of climate change. If the atmosphere that protects us should grow too hot, it would raise an entirely new danger that may raise new issues about humanity's longer-term survival. There is real concern that this heating process, if it should go up on a global average by two or three degrees Celsius, might reach a "tipping point" where reversal of this gradual process might become irreversible. This is, of course, unless some totally new technological solution might be found. Fortunately humans are often clever in finding survival technologies.

Unless one is flying in space above the Van Allen Belts the threats from natural space hazards today remain quite small. These various hazards include so-called solar flares and coronal mass ejections that coincide with the 11-year solar cycle that varies from solar minimum to solar maximum. Most of the times we are quite safe here on Earth, but every 11 years there is a risk that our electrical grids and electronic systems could be zapped big time. We know from The Carrington Event of 1859 and the more recent massive coronal mass ejection of 1989 that these are dangers that cannot be ignored and must be taken seriously [11]. We will also consider the hazards that come from cosmic and solar ultraviolet radiation that is a threat to astronauts and cosmonauts as well as an increasing threat to people in the extreme latitudes near the Polar Regions where the ozone holes now exist.

Chapter 6 addresses the threats posed by solar flares and coronal mass ejections (CMEs). So called "space weather" from the Sun and the cosmos occurs all the time. There are solar eruptions that occur periodically, and during so-called solar max these threatening events are about 15 times more likely to occur than at solar minimum. So-called CME events are characterized by the release of massive amounts of super charged ions that are ejected from the Sun's corona, which is a raging mass of super-heated plasma that reaches one million degrees Celsius. As a result of these periodic solar events a highly destructive mass of ions are released. These ions and charged particles travel at millions of miles an hour and actually pose a major threat to satellites and spacecraft in space. A number of protective measures need to be employed to protect satellites and orbital spacecraft from these occasional blasts, some of which are violent enough to threaten not only not only satellites in orbit but as noted earlier electrical grids, electronic equipment, and facilities on the ground. In short, CMEs, in the most severe cases, can endanger much of the modern infrastructure on Earth. This means not only power grids but pipeline systems and highly distributed computers and telecommunications networks as well. Just think of the consequences if all the microprocessors on all the vehicles and aircraft in the world were to be blown out by a super-massive solar eruption.

Chapter 7 will focus on solar and cosmic radiation and can likewise present hazards to space assets as well as people right here on Planet Earth as well. Widening holes in the ozone layer allow through truly harmful X-ray radiation in the Polar Regions. Solar and cosmic ultraviolet radiation travels essentially at the speed of light or close to 300,000 km/second or 186,000 miles/second. Solar eruptions that contain super charged electron ions as well as alpha and beta particles travel at huge velocities. Despite this great speed these eruptions nevertheless travel on the order of a 100 times slower than the speed of light or energetic gamma rays. This is indeed fortunate. The speed differential allows solar flares and CMEs to be detected via solar observatories and space-based sensors so that satellites and key facilities can be powered down and electrical systems switched off to protect against the "big hits" from these solar storms or super space weather events. Without this type of warning system hundreds of orbiting spacecraft worth hundreds of billions of dollars could be at risk and essential satellite operations lost for communications, navigation, remote sensing, weather forecasting, and military-related services.

Chapter 8 examines how potentially hazardous asteroids (PHAs) and comets pose an ongoing risk to humans, and Chap. 9 addresses what is currently being done to address and forestall these potentially calamitous events. These NEOs are rarely of large enough size to actually pose a major threat, but on average—about every 50–100 million years—these natural orbital debris can truly clobber Earth and its inhabitants. The good news is that we believe that we have identified some 90 % of the potentially hazardous asteroids that are 1,000 m or more in diameter and might come within 9 million miles (or 14.4 million km) of Earth. The bad news is that it is estimated that there are another 10 % of these large threats still to be identified and some 80 % of such asteroids some 100–1,000 m in size to be cataloged. An asteroid of this smaller size could still hit us with the force of tens of thousands of atomic bombs. What is perhaps most important of all is to understand that impacts of objects in this size range are much more frequent than every million years. In fact the chance of a Tunguska-size impact this century is in the order of 1 in 10 to 1 in 5. Later in this book we address the so-called Torino Scale, that is sort of like the Richter Scale for potentially hazardous asteroids. This chart indicates both the likelihood of strikes and the type of damage various-sized NEOs might cause if they hit Earth.

And there are also a large number of potentially hazard comets still to be detected as well. Currently the odds seem to be in our favor, but there are a number of specific asteroids we are tracking with particular concern.

In the short term a much more serious threat for spacecraft are the millions of meteorites and micro-meteorites that can strike and disable a spacecraft. There are a series of recurring meteor showers that pose high levels of risk, but damage from a meteor or even a meteorite can occur at any time. Indeed it is estimated that about 15 % of the strikes on satellites today are from micro-meteorites and not miniscule space junk.

Chapter 10 recaps the major points from the book. Thus this chapter seeks to provide a synoptic overview of the various types of space threats to space assets and even to people residing on Earth or flying through Earth's atmosphere. The strategies and technologies that address these various hazards are summarized as the "Top Ten Things to Know about Space Threats".

Chapter 2
The Space Debris Threat and the Kessler Syndrome

The most beautiful thing we can experience is the mysterious.
It is the source of all true art and science.

–Albert Einstein

Why is the Problem Getting Worse?

One might logically ask this question. If we typically have less than a hundred launches into space each year—after discounting suborbital flights and sounding rockets—why can't we quickly bring the problem of space debris under rather quick control now that we have international guidelines in place? This is not really a mysterious problem, but it is certainly a complex one. The simplest answer is that debris begets debris.

There is a perhaps a somewhat useful metaphor here, which might be helpful to set the problem in context. Although this is certainly not a completely accurate picture it might help to visualize the problem and set the issues of orbital debris clean up in context.

It is not hard to shoot out a large number of street lights, but it can take a long time to clean up the broken glass, repair the sockets and wiring, and restore that which was rapidly destroyed. Further the streetlight, when first installed, consists of a lamp pole, a light bulb and a glass lamp cover. The streetlight that is destroyed may involve hundreds of pieces of debris to be cleaned up and carefully disposed. If just one light were to be shot out in outer space, the pieces would over time spread out over a huge area that would eventually encircle the entire planet. As a thought experiment think now what if one had allowed this sort of damage to continue in this manner for a half century many thousands of times with no effective repairs. It should be clear that a quick clean up and recovery may take quite a while to complete.

J. N. Pelton, *Space Debris and Other Threats from Outer Space*,
SpringerBriefs in Space Development, DOI: 10.1007/978-1-4614-6714-4_2,
© Joseph N. Pelton 2013

The other thing to consider is that if two largest items collide in space at about 25,000 km an hour, the result is not four or six debris items, but perhaps more like 3,000 tracked objects and many more thousands of smaller, untrackable objects.

The collision in this case is more like an atomic explosion in terms of energy release than a stick of dynamite exploding. This type of incredibly high speed crash not only generates a huge new amount of debris elements, but the debris elements over time tend to widely disperse. Figure 1.4 above indicates the dispersal of the 3,000 debris elements from the missile destruction of the International Space Station which is clearly imperilled by this debris. We sincerely need to hope that no more such large collision events occur before we find a way of removing large debris elements from orbit to illustrate the point. The thin white line represents the orbit.

The very careful and rigorous study by J.-C. Liou and Nicolas Johnson indicated in 2006 that just the current amount of debris could generate a tripling of the space junk over the next 200 years. This is because space debris collides and generates more debris of smaller and smaller size. Since Liou's and Johnson's analysis there have been over 500 additional launches, and many of these had multiple payloads. The main problem is thus not cleaning up after new launches (although this is certainly part of the equation) but rather dealing with the current debris that is slowly grinding out additional debris elements. Even here there is a need for "triage" to address the most crucial problem first and then seek solutions to the rest of the problem later. This most urgent part of the debris mitigation process would be to remove from the low Earth Sun synchronous polar orbits the largest pieces of debris first. This is because these derelicts in space could generate the largest amounts of major new debris elements if there would be a major collision. This we know directly from experience.

There have been a number of studies conducted by various space agencies about space debris and its future potential increase. On one hand these studies are reassuring and on the other quite disturbing. At one level, these studies confirm there is a huge amount of open space around Earth relatively free of debris. Even in a so-called "congested area" such as the polar region in low Earth orbit, as depicted in Fig. 2.1, the likelihood of a collision remains extremely small. Figure 2.1 seems so frightening in large part because the scale depicted in this graphic is about 90 million to 1. The worst news of all is that more debris is forming than is returning to Earth due to gravitational effects. In fact there are now well over 6,000 tons of debris in orbit.

Space Debris in Orbit

The creation of additional space debris comes from a great variety of sources such as explosions of fuel tanks, launch vehicle upper stages and fairings as well as active and defunct satellites being bombarded by debris, and so on. Further micro-meteorites from space are constantly raining down on the inner parts of the Solar System. These micro-meteorites are responsible for an estimated 12–15 % of the

Growth of the Satellite Population

National Aeronautics and Space Administration

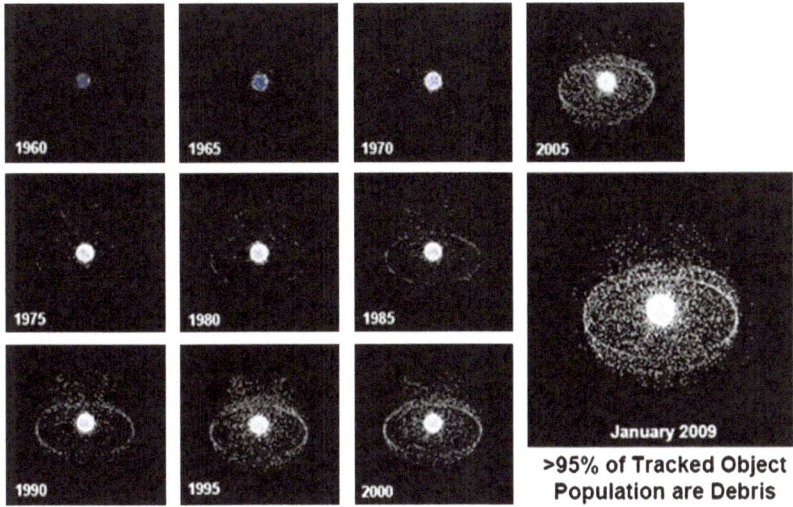

Fig. 2.1 Charting the significant increase in space debris that has occurred since 1960 (Graphic courtesy of NASA)

"hits" on spacecraft, based on the latest studies by various research institutes and researchers that monitor this activity.

Twenty-five years ago the cascade effect of debris crashing into other orbital objects produced a modest amount of new debris elements as can be seen in Fig. 2.1. But in time things began to change. Today this cascade effect is the largest source of new debris elements as the number of micro-debris elements that are less than 1 mm in size has climbed into the millions. There are perhaps enough of these various debris elements from the smallest chips of paint to the largest derelict satellites and upper stage rockets to increase the "number" of debris elements by a factor of four to six times over the next two centuries, even if there was to be a total moratorium on all future launches. This projection is based on the findings from the Liou and Johnson study in 2006 and factors in the number of new elements since that time including the Iridium-Kosmos collision and the Chinese anti-satellite missile test.

Orbital debris is not evenly distributed around Earth's orbit. There are particular bands where these orbital debris are currently concentrated. The worst congestion is in the LEO region and particularly the Sun-synchronous polar orbits. The depiction of the LEO region that is below the Van Allen Radiation Belt is clearly shown in Fig. 2.1 above. The other orbital region such as the MEO region above the Van Allen Belts and the GEO region still contain a number of satellites and debris elements, but relative speaking these are much less congested. This is because that not only are there far fewer debris elements, but also because the debris has a much larger volume in which to spread. Figure 2.1 shows the build up over time of the debris around Earth and how it has escalated in recent years. In 1980 the problem was hardly apparent, and even by 1985 it seemed almost trivial, but today it is clearly a larger and growing issue.

There are a number of other important aspects to note with regard to the orbits that are of importance. One aspect is that there are different disposal concepts that apply to these three different orbits. One logical disposal mode is to fire jets so that a satellite in LEO will simply de-orbit and burn up on its descent or splash down into the ocean. For geosynchronous satellites the disposal method is to push the spacecraft to a graveyard orbit that is higher than the geo orbit. When thus positioned there, these satellites can stay in super synchronous orbit for millions of years.

The greatest challenge is presented by the MEOs in terms of the disposal of satellites at end of life. Only a small amount of increment fuel is required to de-orbit a LEO satellite or to push a GEO satellite into a higher graveyard orbit. The disposal of MEO satellites is a problem in that a 40 % greater amount of fuel—beyond that used for orbital positioning—is needed to de-orbit a spacecraft launched into this orbit. This constitutes a very large economic penalty in terms of launch costs and increasing the size of propellant fuel tanks. Other options might be explored to move MEO satellites at end of life into some type of "graveyard or parking orbit", but this would not be an easy or permanent solution because the Sun, Earth and Moon would impact this type of orbit and thus it would not be stable.

Actually the problem and complexity of this final disposal issue only increases when probed further. Satellites can lose their ability to be commanded and thus be stranded in their orbits. Elements of the launch such as the upper stage rocket, fairings that served to protect the satellite from the atmosphere during launch and other extraneous parts can be launched and stranded in orbit with no mechanism to de-orbit them except for gravitational pull and atmospheric drag. Some satellite operators have claimed that they were requested not to de-orbit their failed satellites from operators of defense-related satellites because of possible collision with clandestine satellites uses for surveillance.

If the launch of a spacecraft is into LEO, these elements will eventually degrade, but this is not the case with MEO or GEO orbit. And, of course, not all satellites are launched into LEO, MEO or GEO orbit. Some satellites are launched into highly elliptical orbits such as the so-called "Molniya" orbit, named after the Russian satellite with this name. Or satellites can be launched into a somewhat similar highly elliptical "Loopus" orbit. There are also various other orbits such as the Quasi-Zenith or Figure 8 orbit (i.e., a geo orbit inclined 45 degrees), super-synchronous orbits and even unintended orbits. These last can result from a launch failure when the rocket fires too long or not enough, and thus the rocket is put on the wrong path.

Once a satellite or rocket motor becomes stranded in orbit, it can become a source of additional debris. Any of these stranded or even actively controlled space objects can be hit by another piece of debris at high speed and generate other debris. A fuel tank or a battery might explode and create additional elements of debris. The recommended procedure of venting fuel tanks for end of life satellites is considered an important procedure now widely practiced to help minimize space debris.

The uneven distribution of orbital debris creates problems with regard to those who perceive this as a serious issue and those willing to support in an active way the cleaning up of the mounting amounts of space junk. Those who operate satellites in GEO orbit are inclined to respond to concerns about rising debris by saying this is largely a LEO and polar orbit problem and not one that affects me. Those who operate MEO orbit systems might say much the same.

The increasing build up over time in orbital debris will, of course, be a problem for everyone who seeks to safety launch into orbit since launchers must travel through LEO on their way to a higher orbit. Further there is increasing debris in all orbits and unless the problem is addressed in the nearer term the longer term costs and difficulty of debris removal will exponential increase over time. Just as the issue of the sustainability of the Earth's environment is becoming exponentially worse and more difficult and expensive to address over time, the same type of problems exist with "kicking the can down the road" with regard to space debris.

The Urgency of Action and Orbital Priorities

The urgency of addressing the space debris problem is clearly perceived to be at different levels by those whose missions are related to LEO, MEO and GEO orbit systems. Thus if there are financial approaches devised to collect funds to address this problem, it is likely that contribution levels might well be different for those launching to LEO, MEO, GEO or points beyond. The discussion of orbital debris also often focuses on which countries are responsible for the creation of this problem in the first place. Clearly it is only a few spacefaring nations that were the prime cause of today's space junk. The primary countries in this regard are clearly the United States, China and Russia.

Although these three countries—or enterprises based in these countries—are the clear source of this debris, the source of secondary, tertiary, or even quandary debris that has come from subsequent collisions in space is much harder to assess. Instead of trying to assign specific responsibility to a particular country and thus looking backwards in time for a solution, it might be more appropriate to try to look forward to a more integrated global solution. The number of countries launching rockets and spacecraft into space is still only ten in number. The three primary launching countries plus Europe launch about 90 % of all rockets into space and well over 95 % of the total payload mass to orbit—and this will likely remain the case for some time to come.

Upgrading Debris Tracking Capabilities

A great deal of activity is now devoted to tracking space debris. Since 1961 the U. S. Air Force has been operating the Space Surveillance System that has been using increasingly outmoded Very High Frequency (VHF) radar tracking and in-orbit resources to track the mounting amount of space debris. As the amount and number of debris objects has increased exponentially, this system has become increasingly unable to keep up with the tracking requirements. This system, which was originally conceived as a means to detect a missile launch attack against the United States, is increasingly utilized to help protect key U. S. orbital assets. This includes anticipating possible collisions with the International Space Station (ISS) by a major debris element and indicating how raising the ISS orbit at the correct time could eliminate such risks.

The U.S. Air Force has contracted with Lockheed Martin to upgrade the existing radar systems and implement what is known as the "space fence" to have much more precise tracking capabilities. The first elements of this new capability were tested in February and March 2012 and successfully demonstrated orbital debris tracking capability. Based on these tests, the air force approved the design and an implementation plan. Steve Bruce, vice president of the Space Fence program for Lockheed Martin, said in a statement after the tests: "Our final system design incorporates scalable, solid-state S-band radar, with a higher wavelength frequency radar capable of detecting much smaller objects than the Air Force's current system [12]." This new space fence system will thus eventually be able to track object in LEO down 1 cm or 0.4 inches in diameter. This is more or less equivalent to the capability to track some 500,000 space debris elements [13].

The control center for this new Mark II space fence orbital tracking system is now operational, even though it will be several years before the new multi-billion dollar capability is fully installed and operational. (See Fig. 2.2)

Fig. 2.2 Mark II S-band radar space fence operations center (Graphic courtesy of Lockheed Martin Corporation)

Space Traffic Management

The space launch environment has clearly become more complex, with growing space launching capabilities and different sorts of commercial space activities. One thought that has arisen with the growth of interest in commercial spaceflight is that of Space Traffic Management. Today commercial spaceflight includes so-called space tourism, commercial cargo and human flight to orbit, commercial space stations and possibly hypersonic transportation systems. This diversity of activity and the increasing "mixture" of aviation, aerospace and space transportation systems suggest that public safety on the ground, in aviation space, in stratospheric operations and in outer space may only be systematically achieved through Space Traffic Management. The relevance here is that if an international body such as the International Civil Aviation Organization (ICAO) is charged with this responsibility, then they might also be able to help oversee this emerging complex aviation/aerospace/outer space environment and also administer processes to control space debris.

In this new role the newly designated international mechanism for Space Traffic Management might set safety and operational standards for many space and stratospheric missions and activities. This agency might coordinate international standards for spaceports, for hypersonic transportation systems and for commercial sub-orbital flights associated with "space tourism," for maximum altitudes for "cubesats," nanosats and microsatellites, for active de-orbit of satellites and upper stage rockets, etc. Even prior to the agreement on this new regulatory regime, which might take many years to achieve, there might be an international code of conduct for space that might suggest better practices, safety standards and debris mitigation standards than exist today.

The Next Steps Forward

The following chapters will address actions that might be taken to mitigate the further increase of orbital debris and processes that might be employed to remove debris from orbit. These mitigation efforts may involve legal, technical, operational or financial steps that can either help to stop the creation of new debris or carry out active removal of debris from orbit.

Chapter 3
Different Approaches to the Space Debris Problem

> *It is immoral to design a product or system for mankind without recognition and evaluation of the hazards associated with that product or system.*

Anonymous Space Safety Engineer

Institutional Arrangements to Address Space Debris

The many challenges that orbital debris presents are daunting. Conventional rocket launches can easily contribute to the debris problem and often do. There are many things that can go wrong in carrying out the active de-orbit of satellites and/or upper rocket stages. Removal of debris through known techniques today is expensive, difficult, and complicated by legal liability provisions of the "Liability Treaty". The debris that is in orbit begets additional debris by the cascade effect of debris collisions. (This process has now created a swarm of millions of micro-debris elements, especially in LEO.) Collision of large debris elements (or even active satellites) generates a very large number of new debris. Using armor on active satellites against debris or micro-meteorites works only up to objects that are 1 cm or less in size. In short there is no magic elixir or wand to wave to make this increasingly serious problem to go away.

Yet new institutional and regulatory approaches to space debris are being actively explored. In light of the fact there are only ten spacefaring nations and most debris can be directly linked to the United States, Russia and China, one would have thought more progress would have been achieved on this front than is the case today. Certainly space debris issues are being actively addressed in a number of ways, such as:

- Prelaunch due diligence.
- Improved operational procedures, including controlled de-orbiting of spacecraft, venting of toxic gases prior to de-orbit, and avoiding the use of nuclear-powered systems for satellites that are to eventually de-orbit back to Earth.

J. N. Pelton, *Space Debris and Other Threats from Outer Space*,
SpringerBriefs in Space Development, DOI: 10.1007/978-1-4614-6714-4_3,

- Development of new technologies that might be able to achieve space debris removal or to actively avert major space debris elements from colliding, or to provide more shielding.
- New legal procedures and agreements to address among other things space debris matters. (There is a growing need to address the troublesome problem of national liability with regard to launched "space objects" that is actually serving to retard efforts to undertake active debris removal.)
- New processes and mechanisms such as a fund for debris removal or a new institutional mechanism to address space debris removal.

Yet much more needs to be done. There are now no significant or specific penalties that apply to creating new space debris except a general liability provision that says that if your "space object" injures someone else in another country then you are at fault and subject to a liability claim. In short one of the big problems is that there are no truly effective incentives or penalties that would "encourage" countries to stop creating new debris or remove debris from orbit other than a good public image and not being seen as a "bad guy".

Most articles written about space debris tend to focus on either the orbital mechanics, the space technology needed to remove debris from orbit, or relevant regulatory issues. Technical papers usually seek to address such aspects as: (i) the growing extent of the problem and space situational awareness; (ii) the factors that are contributing to the rate of buildup of debris; or (iii) technical approaches related to debris removal and remediation. Regulatory papers on the other hand tend to address: (i) various ways to undertake due diligence to prevent the creation of new debris; (ii) the actions needed to be taken by governmental or intergovernmental bodies relating to orbital debris and its mitigation; (iii) questions of liability and legal responsibility; or (iv) the creation of agencies or mechanisms to undertake space situational awareness, to control debris and/or to remove debris from orbit.

A Global Fund for Debris Removal?

The missing element in many of these discussions is how to create the economic wherewithal to address the debris problem and how to create financial incentives to correct the problem. In this section the analysis is directed toward the merits of establishing national, regional and in time perhaps universal agreements to establish economic funds—as well as incentives or penalties—to mitigate the problem. The purpose of such funds would be several fold: (i) to create a rebate system to reward "clean and debris free" launches; (ii) to award a further rebate to reward clean disposal of satellites at the end-of life. Under this approach there would now be clear incentives to get rid of space debris as opposed to the current disincentives and potential liabilities associated with bringing debris and satellites down or into graveyard orbits. The creation of a fund—or perhaps several funds that could grow into a global fund—would create incentives to develop the best technology rather than a single approach that might ultimately prove to be suboptimal.

The 20-year sunset for the fund(s) would create a specific goal to complete the mission, and if success is achieved there would not be the additional issue of having to disband an international agency.

The fund (or collection of national/regional funds) could be established over time in an "organic manner" with countries forming such a fund on a national basis, or perhaps Europe could form such a fund on a regional basis. This type of national, regional, and in time ultimately universal fund would be formed by space actors for the specific purpose of addressing the space debris issue. This approach would thus become a pro-active "forward looking" approach to financing a solution to the problem rather than seeking a "backwards-looking" approach to addressing space debris with no financing mechanism in place and nations being "coerced" into doing the "right thing".

The money to capitalize this type of space debris fund would be collected prior to all launches and would equivalent to perhaps 3–5 % of the total cost of various space-related missions. Under this approach LEO/polar orbit missions might be required to pay in 5 % of mission costs. MEO and GEO orbit and deep space missions might be asked to pay in a lower amount. This fund would be collected for a period of perhaps 20 years but would have a sunset provision on the premise that migitation of orbital debris could be successfully accomplished over this length of time. Thus there would need to be an active agreement to extend the fund or it would otherwise elapse.

Such a fund (or network of funds) would be formed by means of a specific assessment paid into a designated bank account (or space insurance company) prior to launch. This fund would apply to all those deploying spacecraft into Earth orbit, or, if on a national or regional basis, would apply to all launches from that country or region. Organizations launching satellites beyond Earth orbit would also pay into the fund but a lower amount. After each launch there would be a partial rebate, assuming it was a certified as a clean "debris-free" launch as independently verified. When a spacecraft was de-orbited at end of life or successfully placed in a graveyard orbit there would be a further rebate. The size of the "clean launch" and "successful disposal" rebates would be specified at the time the fund(s) were established. Approximately half of the payments into the fund, however, would always be retained to compensate those entities involved in removing "officially designated" debris from orbit or moving defunct space objects to a graveyard orbit.

The prime purpose of the national, regional or hopefully, global space debris fund would be to compensate those entities "licensed under an appropriate regulatory framework" to remove debris from Earth orbit or those that develop and operate systems to avoid collisions. This licensing process for entities designated to undertake orbit debris removal or collision avoidance activities might, for example, be formally assigned to the United Nations Office of Outer Space Affairs or in time spelled out in a new international space convention.

Other entities might also be "licensed" by the U. N. Office of Outer Space Affairs to undertake activities associated with the prevention of space debris or space debris mediation or collision avoidance activities separate from the active removal of space debris from orbit. Such activities, however, would be limited to no more than a set percentage of the available funds.

Payment into this fund would "seem and feel" to satellite operators and governmental space agencies conducting space operations very much like buying launch insurance for a spacecraft mission. Indeed the fund could possibly be administered by launch insurance companies. These payments would be different in that it would only represent about a third of the cost associated with purchasing launch insurance, and rebates would eventually return half of the money paid into the fund. Further, the projected end date for the fund would establish a very real goal for accomplishing "a largely space debris-free world". The creation of this fund and the rebate payments would reverse the current incentives that actually "encourage" the increase of orbital debris. Under current space law the owners and operators of space objects not only lack an incentive to remove their space debris from orbit; they actually face substantial financial penalties if the removal process somehow adversely affects another space object and creates liabilities which they are compelled to pay.

The payments into the fund are actually modest when compared to the damages that will ensue once we reach the Kessler syndrome stage and debris continues to cascade out of control on an exponentially increasing basis. Indeed payments for launch insurance operations over the last three decades have varied from a low of about 6 % of total mission costs to as much as 20 % of total costs. Today typically 15 % of mission costs is for launch insurance. If one considers this wide range of payments for launch insurance and the importance of the long term sustainability of space and safe space access one should consider a 5 % orbital debris fund as not being at all excessive or unreasonable, especially if half of the money is ultimately rebated in the advent of a "clean" launch with upper stage rocket motors and launcher fairings being removed from orbit and the satellite eventually disposed of as well.

There would appear to be merit to a flexible "economic fund" approach as opposed to seeking to create a single international agency charged with space debris remediation that would likely focus on a preferred technology and a single approach to debris removal. Licensed international entities, under the fund approach, would not be restricted to a single country. Each country or region that acted first to create orbital debris funds could also give research grants to entities embarked on developing new technology to remove debris from orbit with the latest technology.

In short it is believed that there would be "economic and political efficiency" in having a number of licensed commercial entities capable of developing a diversity of innovative technologies to carry out space debris removal. Overall it is believed that the "economic fund" mechanism could help to create all the right incentives: (a) to reward entities for a clean launch of the satellite and removal of upper stage rockets and protective fairing covers from orbit; (b) to reward operators for removing debris properly at end of life; (c) using the "sunset provision" to establish a specific goal to get the job done; (d) using the "fund approach" (or alternatively even a prize approach) that would allow the competitive development of the best and most cost efficient technology and (e) there would be no need to "dismantle" an international agency at the end of the process.

The Economic, Social, and Strategic Importance of Space

Space applications have become a very diverse and increasingly important aspect of our global society. Over time space applications have expanded in scope, divided into many submarkets, and have evolved into a series of many different "space actors". These include civil governmental space agencies, defense-related space agencies, commercial launch operators, operators of various commercial spacecraft organization, and even public service space operators that are operated by both commercial and non-profit organizations.

The various governmental, defense and commercial space markets are today quite large, with all related annual space applications expenditures and revenues and expenditures totally perhaps $300 billion (U.S.), that would be more or less evenly divided between commercial and military/strategic/governmental programs. The true impact of space activities is not simply a function of their economic size, however, but rather their overall impact on society. Space-related activities today relate to national security, the monitoring of possible attacks via nuclear-armed missiles, the use of space navigation to control transportation (including the takeoff and landing of aircraft), the deployment of satellites for voice, data and television communications, and the use of satellites to forecast weather and avert the impact of hurricanes, typhoons, and other violent weather.

There are many remote sensing operations that observe Earth to detect natural resources, conduct fishing operations, to monitor oil spills, and chart the impacts of climate change. These remote sensing activities have a variety of public service and commercial goals, as is the case with telecommunications satellites, navigation satellites, and other types of application satellites. If humans were to suddenly lose all of its civilian and military communications satellite systems, remote sensing, weather and space navigation systems, the modern "Western world" as well as many other countries would find itself paralyzed. In many ways the result would be like a massive global electrical power failure.

If one thus views the almost overriding importance of sustainable access to space for many centuries to come, the idea of a global fund for active debris removal (ADR) and mitigation almost seems to be a quite modest proposal. Commercial organizations willingly spend 15 % of their net investment on a new satellite network and its positioning in space for launch insurance, if not more. To have the ten governments of space-launching nations agree on a "space debris removal mitigation fund" that would require a much lesser amount of money than that spent on launch insurance does not seem unreasonable.

Further if there were rebates for "clean launches" and additional refunds after satellites were de-orbited this would serve several additional positive purposes. It would make the system more equitable by providing partial refunds to those whose actions minimized the future formation of debris. It would reward responsible action and in effect also serve as a fine against those who did not act responsibly. It would help finance future mitigation actions to achieve future "debris free" launches and also assist in funding future efforts to remove space debris. And finally the existence of the fund would allow flexibility to back alternative removal technologies and also the shut down of the fund once near Earth space was indeed cleaned up [14].

A New International Arrangement for Space Debris Removal?

Other alternative concepts to address the mounting problem of space debris have also been proposed. One concept developed in another International Association for the Advancement of Space Safety (IAASS) study called for another mechanism for assured debris removal, an integrated development, operational and regulatory framework for space debris akin to an inter-governmental organization modeled on the intergovernmental agreement under which the early INTELSAT organization was created. This study envisioned the creation of INREMSAT (for International Removal, Maintenance and Servicing of Satellites).

Under this scenario INREMSAT-subscribing governments would be asked to sign a legal instrument (i.e., either a treaty or international agreement) to procure INREMSAT's services for the removal of a number of existing "big" space debris (dead satellites, last stage rockets, rocket fairings used to shield satellites during launch, etc.) from Earth orbit. In addition the ten spacefaring countries would be required to make changes to their national space licensing rules to include an "assured removal" clause as a condition for obtaining a license to launch and operate a satellite regardless of whether they were to employ a national or foreign launch service. Such a clause would apply not only to the satellite but the upper stage of the launcher and its fairings.

The "assured removal" clause would further require that the operator demonstrate, subject to independent verification, that the satellite network in question had the capability (and specific plans) to perform autonomously at the end-of-life a safe controlled re-entry or it could be removed to a graveyard orbit. This study proposed that national governments, military systems and/or commercial organizations would contract with INREMSAT or a similar commercial service provider for such activity. Finally, the operator under this proposal would to commit to purchase an insurance policy in case a failure or malfunction occurred. In this case the insurance company would then procure and cover all costs of the relevant disposal service [15].

Regardless of whether a global debris fund or whether an international agreement or treaty along the lines of INREMSAT is implemented, the need for new international agreements and mechanisms are clearly needed to alleviate the escalating buildup of space debris that is clearly depicted in Fig. 2.1 in Chap. 2.

Chapter 4
Who is Addressing Orbital Debris Problems?

> *It is time for a new generation of leadership to cope with new problems and new opportunities. For there is a new world to be won.*
> –John F. Kennedy, July 4, (1960)

Key International Organizations

The two most important international coordinative bodies that are addressing the space debris problem are the Inter Agency Space Debris Coordinating Committee (IADC) and U. N.'s Committee on the Peaceful Uses of Outer Space (COPUOS). The IADC includes most of the world's space agencies with a space launch capability while the COPUOS now includes some 70 nations and addresses a wide range of space-related issues including that of mitigation of space debris.

Inter-Agency Space Debris Coordination Committee

The IADC is an international governmental forum of space agencies who are engaged in worldwide coordination of their activities in order to seek to mitigate and minimize the adverse affects of man-made and natural debris in space.

The primary purposes of the IADC include: (i) the exchange of information on space debris research activities between member space agencies; (ii) to facilitate opportunities for cooperation in space debris research; (iii) to review the progress of ongoing cooperative activities; (iv) and to identify debris mitigation options.

The IADC member agencies include the following:

- ASI (Agenzia Spaziale Italiana)
- CNES (Centre National d'Etudes Spatiales)

J. N. Pelton, *Space Debris and Other Threats from Outer Space*,
SpringerBriefs in Space Development, DOI: 10.1007/978-1-4614-6714-4_4,
© Joseph N. Pelton 2013

- CNSA (China National Space Administration)
- CSA (Canadian Space Agency)
- DLR (German Aerospace Center)
- ESA (European Space Agency)
- ISRO (Indian Space Research Organization)
- JAXA (Japan Aerospace Exploration Agency)
- NASA (National Aeronautics and Space Administration)
- NSAU (National Space Agency of Ukraine)
- ROSCOSMOS (Russian Federal Space Agency)
- UK Space (UK Space Agency)

The IADC includes a Steering Group plus four Working Groups. These Working Groups include: (i) Working Group 1 on Measurements; (ii) Working Group 2 on Environment and Database; (iii) Working Group 3 on Protection; and (iv) Working Group 4 on Mitigation.

These various integrated parts make up the IADC effort. It was the IADC that after extensive international collaboration managed to develop guidelines for countries that would reduce the formation of future space debris [16].

In July 2007 the IADC adopted a common set of standards for addressing space debris that focused on four important areas. These standards thus addressed:

- limitation of debris released during normal operations;
- minimization of the potential for on-orbit break-ups;
- post-mission disposal;
- prevention of on-orbit collisions.

These guidelines also usefully served to provide a specific, internationally agreed definition of space debris as follows: "Space debris are all man-made objects including fragments, and elements thereof, in Earth orbit, or re-entering the Earth's atmosphere, that are non functional [17]". Such a clear and sweeping definition is of significance, because none of the U. N.'s approved treaties or international agreements or conventions refer to anything other than to "space objects".

The U. N. Conventions and COPUOS

The current U. N. conventions—particularly in the context of the now crucial liability convention—make no distinction between functional and non functional spacecraft or launch vehicles. The lack of such a distinction and the lack of an internationally agreed upon process of naming and recognizing space debris have served to undercut the incentive for nations to remove their own space junk from orbit [17].

The IADC, working in cooperation the U. N.'s COPUOS, has been seeking to develop procedures to mitigate space debris for almost two decades. After 18 years of deliberations COPUOS agreed to adopt a detailed set of voluntary guidelines

Fig. 4.1 The U. N.'s Vienna
International Center Meeting
Site for COPUOS (Graphics
courtesy the U.N.'s Office of
Outer Space Affairs)

for all spacefaring nations that were based on the IADC deliberations. Despite these quite useful efforts by the COPUOS the problem persists (Fig. 4.1).

The Space Data Association

In parallel the commercial space industry has actively sought to find ways to mini-mize the chance of satellites crashing into one another. This initiative is known as the Space Data Association, which was formed on the initiative of Intelsat General, SES of Luxembourg and Inmarsat. Although this initiative started with some of the leading members of the satellite communications industry with assets in geosynchronous orbit, it has continued to expand to include not only satellite communications operators in various orbits but also operators of meteorological and remote-sensing systems. The computer-to-computer real time updating of sys-tem operators allows those with very valuable assets in orbit to know of possible conjunction of satellite that might be at risk for an in-orbit collision. The software that allows real-time data exchange is provided by Analytic Graphics Inc. (AGI), and the corporate management is provided by Mansat of the Isle of Man [18].

The current membership of the non-profit consortium now includes:

- Amos by Spacecom
- Arabsat
- Avanti of the United Kingdom
- Echostar
- Eutelsat
- Geo Eye
- GE Satellite
- Inmarsat (Executive Member)
- Intelsat (Executive Member)
- National Ocean and Atmospheric Administration of the U.S.
- Paradigm (Prime contractor for the U.K. Skynet system)
- SES Astra (Executive Member)
- Space Systems/Loral

- Star One (a company of Embratel)
- Telesat

The membership has continued to grow and in time may include most operators of application satellites around the world [18].

The Way Forward

The progress made by the IADC, COPUOS, SDA and the expanded space-tracking capabilities being established to track space debris more accurately and in smaller elements all represent steps forward. Most spacefaring nations are observing the voluntary guidelines to minimize space debris. Clearly additional steps are needed such as to create national, regional or global funds for space debris removal, a new international agency to police debris minimization and to undertake debris removal, and modification of the U. N. Space Convention on Liability. Individual nations on their own initiative might undertake to remove large debris elements from space, but this seems less likely to happen until some new institutional arrangements are put in place. One can only hope that the COPUOS Working Group on the Long Term Sustainability of Space can provide some impetus forward, but it is still too early to predict what the outcome from this initiative will finally be.

Chapter 5
Technological Approaches to Debris Removal or Mitigation

It is important for the Human Race to spread out into space for the survival of the species.

—Stephen Hawking

Developing the Technology to Eliminate Space Junk

There is a growing consensus view among experts that an active process to remove unwanted orbital debris will be needed to ensure the long-term viability of key orbits and to ensure the long-term sustainability of space. Unless such a process is undertaken there will be a major threat to the use of space around Earth for satellite communications, space navigation, remote sensing, meteorological services, scientific experiments, and habitats with crews on board. A wide variety of techniques to "remediate" (i.e., de-orbit) orbital debris have been suggested, but none have been proven to be technically or economically viable, and many of these techniques could also be considered to be "space weapons".

Some of the more commonly identified techniques now being explored are thus briefly addressed here in terms of technical approach and feasibility, level of readiness, and policy and regulatory issues and concerns.

Electro Dynamic Debris Elimination

Technology

This technical approach is appealing in that it uses Earth's magnetic field for propulsion and thus can operate cost efficiently and over an extended period of time. It is a large-scale (several kilometers in size) mechanism but nevertheless not

J. N. Pelton, *Space Debris and Other Threats from Outer Space*,
SpringerBriefs in Space Development, DOI: 10.1007/978-1-4614-6714-4_5,
© Joseph N. Pelton 2013

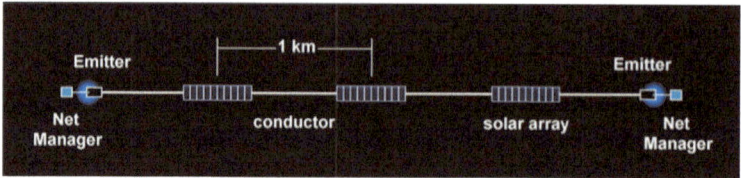

Fig. 5.1 Conceptual diagram of an electro-dynamic debris elimination (EDDE) system

massive in that the elements of this device would be linked together by very light-weight but strong "tape-like" elements. This approach nevertheless still requires a very significant capital investment and an "integrated approach to debris removal". This concept is shown in Fig. 5.1 [19].

Level of Readiness and Feasibility

The physics and theory are clear, but the need for a scale-model demonstration would seem advisable before a full-scale system were launched. This is one of the more mature ideas, and it is being commercialized by funding provided by the U.S. government. It would only be able to address LEO space objects since Earth's magnetic field is not strong enough to address operation in higher orbits.

Policy and Legal Concerns

There would be a number of policy concerns with the highly "integrated and holistic approach" to debris removal. This would likely require a new convention to allow a commercial entity to remove space objects of other launching nations from orbit. The system would need to be designed and operated on an international basis in such a manner that it would not be considered a space weapon.

Tether-Based De-Orbit Systems Using Electro Dynamic Techniques

Technology

This is a much more piece-meal approach than the previous technique. In this case a robotic device would attach a tether to a derelict satellite or debris element. The tether's movement through Earth's magnetic field could potentially power an ion electronic thrust system to aid de-orbit. The asymmetry of the tether would create wobble to help aid de-orbit as well.

Level of Readiness and Feasibility

This approach could be tested more rapidly and a robotic system might be able to attach several tethers to debris objects—either with or without ion engines to assist with de-orbit.

Policy and Legal Concerns

The policy and legal concerns would be much the same as with the EDDE mechanism. Because this mechanism would be on a much smaller scale the concerns with such an approach would be much less.

Much Higher-Powered Ground-Based Lasers

Technology

This approach is similar to the collision avoidance technique but would use higher-power laser systems that would be capable of firing gigawatt pulses at debris [20]. Small debris elements would be moved rather quickly to a new orbit that would decay due to the pull of gravity. In the case of larger derelict satellites, however, it would take a continuing array of pulses to push a large mass into a new orbit that would accomplish de-orbiting.

Level of Readiness and Feasibility

Such high gigawatt GBL systems, although developed in the lab, have not been operationally tested. Such a high-powered laser could clearly be used as a space weapon. A targeting system and actual laser pulsing mechanism needs to be operationally demonstrated to prove that such a system would work. Although this approach would be expensive and has not been operationally demonstrated, it is based on military research and a technique that could likely be implemented in just a few years' times.

Policy and Legal Concerns

A number of countries have strongly objected to such use of ground-based lasers as an anti-satellite weapon that they believe are contrary to Article 4 of the Outer Space Treaty. These countries feel that new international conventions are required on such specific aspects as demilitarization of space, space debris mitigation and remediation, and that action needs to be taken to create an internationally sanctioned entity to address space debris and space traffic management issues.

Solar Sail Devices

Technology

Such solar sail arrays would be designed to attach themselves robotically to large debris object, and over time this would facilitate the active de-orbit of the derelict satellite [20]. This would work only on LEO debris and would not assist with MEO and GEO debris.

Level of Readiness and Feasibility

This approach would require several more years of development work before it could demonstrated as feasible—let alone being operationally implemented. It would thus likely require at least 5 years to implement, and it would be expensive because of the cost of manufacturing the solar sail and associated thrusters and robotic equipment and the launch costs. One solar sail would likely be required for each large debris element removed from orbit.

Policy and Legal Concerns

This approach does not raise specific space weapons concerns, but the international community would still likely wish the solar sail devices to be deployed via a global sanctioned organization than have a particular country (even a civil space agency) carry out these operations.

Tether-Deployed Nets

Technology

This system would deploy "nets" around smaller elements of space debris and speed up their de-orbit (This system has been called "Rustler" for "Round Up of Space Trash—Low Earth orbit Remediation") [20]. The approach is relatively low technology in principle, but this is still more of planning concept than an operational program.

Level of Readiness and Feasibility

Much more R&D is required to develop an operationally viable system, but tether deployment might reduce launch costs.

Policy and Legal Concerns

This approach would probably not raise specific space weapons concerns, but the international community would still likely wish such devices to be deployed via a global sanctioned organization than have a particular country (even a civil space agency) carry out these operations.

Space Mist

Technology

Satellites would be deployed in LEO that could spray gas mists, and the frozen gas mist would serve to bring down small orbital debris elements [20].

Level of Readiness and Feasibility

The concept is well defined, and the key missing element would be an operational test to determine whether the mist would dissipate into space and not achieve the desired result. This technique could, however, be cost effective.

Policy and Legal Concerns

This approach does not raise specific space weapons concerns, but the international community would still likely wish spacecraft that would spray the mist to be deployed via a global sanctioned organization than have a particular country (even a civil space agency) carry out these operations.

Robotic Systems

Technology

Robots would clamp on to space debris and then essentially throw the object into an orbit that would rapid degrade. This concept is similar to that of the solar sail approach [20]. Yet another variation on this theme is known as Slingsat developed at the

University of Texas A &M. This would not only seek to "sling" debris in a path toward de-orbit but use the momentum achieved to maneuver to the next debris element.

Level of Readiness and Feasibility

These various systems vary from quite expensive to the lower-end Slingsat. All require operational testing and certainly robotically-maneuverable are technically demanding. One approach would have the robot self destruct with the derelict spacecraft, while the other would have a series of detachable components that would fly into a de-orbiting path. In November 2010 the Russian Rocket and Space Corporation Energia announced an ambitious plan to build a large nuclear-powered robotically controlled "space pod" that would "knock" derelict satellites out of orbit and would operate over a 15-year lifetime. This technology has also been explored as a way to move the orbits of Earth-threatening meteorites or asteroids [21].

Policy and Legal Concerns

All such systems from the nuclear-powered Russian vehicle down to the slingsat systems might well be considered an anti-satellite weapon, and there would be concerns as to how such a system might be operated and who would control it. None of the proposed systems would be totally free of concerns and thus some agreed level of international control over the operations would likely be necessary to alleviate possible use of such systems as space weapons.

Adhesives

Technology

In this approach very sticky adhesive balls composed of substances such as resins or aerogels would be "shot" on to large space debris so as to alter their orbits and to bring them down over time [21].

Level of Readiness and Feasibility

This system would be less expensive than many of the other approaches but requires actual in-orbit tests.

Policy and Legal Concerns

This approach probably does not raise specific space weapons concerns, but the international community would still likely wish spacecraft that would spray the aerogels to be deployed via a global sanctioned organization than have a particular country (even a civil space agency) carry out these operations.

Ion Beam Shepherd Concepts

Technology

The concept would have a precisely focused hyper-velocity ion beam applied to a piece of space debris that would then "shepherd" the space debris to a controlled de-orbit [22].

Level of Readiness and Feasibility

This is still largely a theoretical concept that has been identified as a possible solution to the space debris problem by JAXA, ESA and academic researchers. It is 3–5 years away from in-orbit tests.

Policy and Legal Concerns

This approach seems to pose few major policy concerns other than the strong need to have international control over these systems and their operation. None of the proposed systems would be totally free of concerns and thus some agreed level of international control over the operations would likely be necessary to alleviate possible use of such systems as space weapons.

Ways to Approach the Solution of the Space Debris Problem

Regulatory Approaches

Clearly any of the various ways forward will include a regulatory component. COPUOS has developed voluntary guidelines for orbital debris control and minimization. These have been developed in conjunction with the IADC that has provided an expert level of technical support. The SDA has also contributed quasi-regulatory concepts to the control of space debris. Nevertheless, there remains a need for better and stricter guidelines to control formation of debris. Certainly the removal of space debris has technical and economical challenges not likely to be solved by regulations alone.

Vetting the Technological Approaches

The current wide diversity of ideas about how to remove space debris suggests several things. It suggests that at this stage at least there is no clear "winning idea" about how this could be done and that all of the options now available are unproven and expensive. Many of the removal concepts give rise to a variety of concerns. The idea that there could be several demonstration projects to start the removal of the largest derelict objects thus makes a great deal of sense. Certainly there are particular concerns about the largest defunct space objects that if they were hit by orbital debris of any significant size could generate hundreds if not thousands of new space debris elements. The Galaxy 15 (pictured in Fig. 5.2 below) is just one of the large derelict space objects that might be considered prime targets for removal from geo orbit in addition to the efforts to remove debris from LEO. In this instance a low-thrust ion engine could accomplish such a mission to move the Galaxy to a graveyard orbit.

The creation of a single international agency to carry out this task gives rise to a host of concerns. These concerns include: (i) the high likelihood of focusing on a single and perhaps ultimately the wrong technology; (ii) extremely high cost; (iii) and the problem of international agencies not necessarily being the best source of innovation, not likely to produce cost effective solutions, and most likely to seek to be self-sustaining even if their mission has been fulfilled. Today the most

Fig. 5.2 The Galaxy 15 satellite now a major piece of orbiting space debris (Graphic courtesy of Boeing)

"competent" space agencies to deal with space situational awareness and space debris removal are most probably military entities, and this gives rise to concerns about space weapons and military uses of systems developed to remove space debris from orbit.

Plan B: Collision Avoidance Systems Using Lasers

There is at this stage considerable skepticism that any of the above technological means will come to maturity and demonstrate operational capability at reasonable expense any time soon. This is why there is considerable attention to what might be considered Plan B. This would be to use lower energy (non-weapons' grade) laser systems to pulse a debris object that is considered likely to collide with another orbital object for long enough to divert the orbit enough to avoid the collision. Since the speeds are so rapid, only a small slowing of the velocity drops the orbit sufficiently to avoid the collision (Fig. 5.3).

Such a process does not require orbital launches and would thus be far less expensive than any of the active removal processes. A review of this process thus reveals the following conclusions.

Technology Approach: This approach would involve the aiming of a focused laser beam toward a satellite or large piece of orbital debris that is determined to be in danger of collision, in order to alter the orbit by lowering it to a lesser altitude. In light of the very high speeds that orbiting objects travel around Earth, only a very small incremental velocity change can serve to avoid an impending collision, especially if the detected possible collision is detected well in advance. This is not of course a solution but a useful mitigation technique that would delay the buildup of new debris and avoid major collisions that could generate thousands of new debris elements.

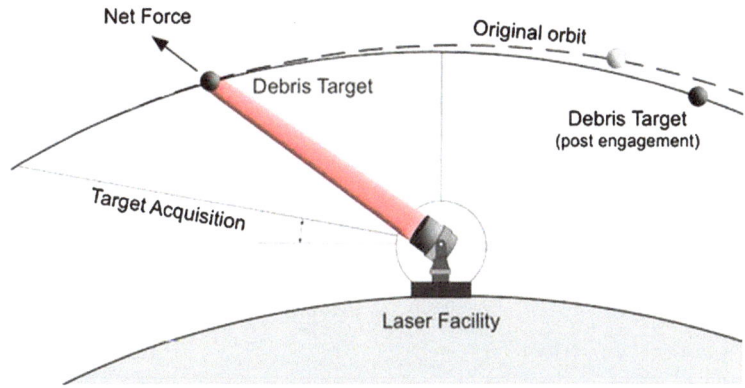

Fig. 5.3 Diversion of orbital objects via laser pulses to avoid a collision

Level of Readiness and Feasibility: This approach is one of the more mature technological concepts and would certainly be the most cost effective in that it would not require the launch and deployment of space assets to cope with the buildup of new debris elements.

Policy and Legal Concerns: The medium intensity laser that would use photons to change a satellite or debris element might still be seen as a weapon. It has been suggested that the nation responsible for the launch of the affected space object could be given operational control of the laser so that officials of the country that had launched the craft or debris element would be responsible for the diversion process.

Chapter 6
The Threat of Space Weather

> *The World, the race, the soul—in space and time the universes,*
> *All bound as benefitting each all—surely going somewhere.*

> –Walt Whitman, On Going Somewhere

Coping with Solar Storms

Solar energetic particles (SEPs) and coronal mass ejections (CMEs) pose a significant threat to in-orbit satellites as well as to Earth's infrastructure in terms of extensive damage that could be caused to the modern electronic grid, computer processes, telecommunications networks, all types of electronic devices (even pipelines) and essentially everything that can conduct electricity. In light of the pervasiveness of electronic devices and electrical systems throughout modern society the risks grow larger each year on an almost exponential basis. A massive solar flare or intensive forms of solar wind can have a devastating effect across the world in terms of automobiles, appliances, computer and telecommunications networks. It is a serious potential problem indeed. The threat extends to aircraft in the skies, electronic grids on the ground and even subterranean networks due to a phenomenon called ground-induced currents (GICs). This chapter discusses the current global monitoring and response systems that address global flares and coronal mass ejections. Today the protective response systems include powering down of spacecraft and switching off their electrical currents when the massive surges come from the Sun, but also Earth-based protective actions to help forestall massive energy failures such as occurred in March 1989 as a result of a large coronal mass ejection (Fig. 6.1).

Coronal mass ejections and super intense space weather have the potential to zap and disable operational satellites. Solar monitoring systems on the ground and now in space constantly monitor the Sun to detect such eruptions so that orbiting satellites can power down and achieve maximum protection against such events,

J. N. Pelton, *Space Debris and Other Threats from Outer Space*,
SpringerBriefs in Space Development, DOI: 10.1007/978-1-4614-6714-4_6,
© Joseph N. Pelton 2013

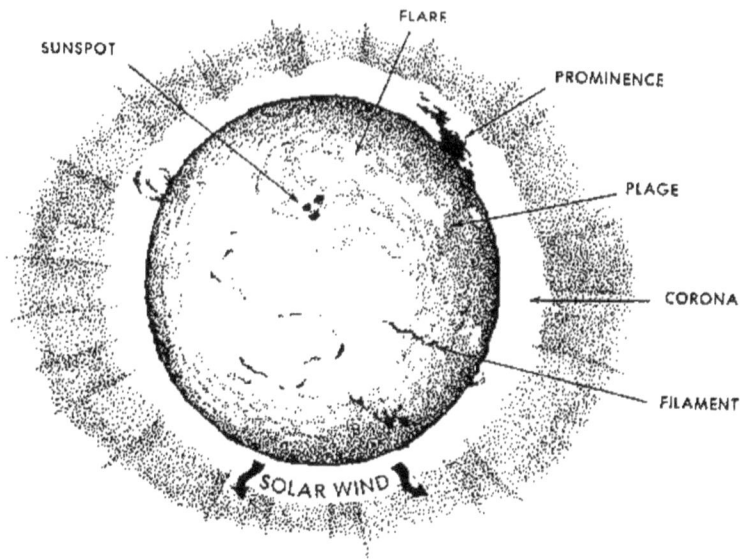

Fig. 6.1 Solar coronal activities that are threats to satellites and to Earth (Graphic courtesy of NASA)

but there is more at risk than just spacecraft. A sufficiently powerful blast from the Sun can also adversely affect electric power grids, computer networks, and more. Figure 6.2 shows a transformer before and after being hit by a major ejection on March13, 1989 [23].

One of the consequences of a coronal mass ejection is a ground-induced current or geo-magnetically induced current (often expressed simply as a GIC). In the case of a very large coronal mass ejection GICs can flow as deep as 20–25 km (12.5–16 miles) within Earth. One such solar event occurred on March 13, 1989. As a result of the CME a number of transformers failed or were rendered useless, including the one shown in Fig. 6.2. The March 1989 coronal mass ejection created a massive power failure from Chicago to Quebec. Overall this event affected many millions of people and lasted for many hours. A strong enough CME can affect power grids at all latitudes, but the ground-induced currents and the most severe impacts on grid systems are most likely to have their strongest impact farther to the south or north [24]. This is because the Van Allen belts tend to divert these incoming surges off to the polar regions. A strong ground-induced current can enter and destroy electrical grids and transformers, but it can also adversely affect pipelines, telecommunications networks, other electrical devices and even hydrocarbon production. Trying to protect critical infrastructure or key aircraft or vehicles from a massive CME would not only involve putting these key assets in underground structures but also ones that are sealed with protective insulation coatings.

Fig. 6.2 Transformer
destroyed by a coronal mass
ejection in 1989 (Graphic
courtesy of NASA Science
News site)

PJM Public Service
Step Up Transformer

Severe internal damage caused by
the space storm of 13 March, 1989

The truth is that the highest southern and northern latitudes are almost daily affected by ground-induced currents that are driven by modest levels of auroral activity. Fortunately the Van Allen belts generally tend to divert incoming alpha and beta particles and gamma rays to the polar regions and protect all forms of life, which are largely concentrated at the low to moderate latitudes. In short Earth is constantly bombarded by solar and cosmic radiation that includes photons and high energy super X-rays (i.e., gamma rays), alpha and beta particles, and other elements of space weather on a 24/7 basis.

In 1997 the Solar and Heliospheric Observatory (SOHO), a joint undertaking of NASA and the European Space Agency (ESA) was launched in order to study the Sun. The mission was designed with a particular focus on a better understanding of the concept of space weather and the powerful coronal mass ejections that occur during the course of the Sun's 11-year cycle that ranges from its lowest to its highest level of activity. This monitoring of the complete cycle is important because of the great variation in solar activity. During solar max, the Sun can have as many as three coronal mass ejections in one day, while during the solar minimum CMEs can be as few as once every four or five days. The reason of this cycle and why the activity can be fifteen times more frequent from solar max to solar minimum is still not understood [25].

Although this mission with joint funding by NASA and ESA was initially scheduled for only a two-year life time, it has now been extended six times. The lifetime of the SOHO extended from 1997 through 2012 and has thus been able to accumulate data over the entire 11-year solar activity cycle. Although this space mission cost a total of about $1.5 billion the costs were divided between the two space agencies to spread the cost of this expensive undertaking, and this mission has produced a great wealth of data.

Some of the key results that have come from SOHO observations include:

- Three-dimensional images of the structure of sunspots below the Sun's outer coronal surface.
- Detail images of the Sun's convection zone, which extends throughout its turbulent outer shell where sunspot activities appear to form.
- A series of solar measurements of the temperature structure, the Sun's processes of interior rotation, and the nature of gas flows below the turbulent and super heated corona area. (This region heats up to the amazing level of 1,000,000 °C.)
- Acceleration rates for both the so-called "slow" and "fast" solar wind.
- A better knowledge of the source and acceleration mechanisms for the "fast" solar wind in what are now seen to be the magnetically "open" regions at the Sun's poles.
- Identification of coronal waves and solar tornadoes.
- An increased ability to forecast space weather and in some cases a new ability to give up to three days notice of major new coronal mass ejections.
- Creation of new early warning system for space weather [26].

More recently in 2006 NASA launched two satellites—one ahead of Earth in solar orbit and one behind Earth so as to be able to witness the characteristics, speed, dimensions and intensity of coronal mass ejections of the Sun. These solar observing satellites are aptly know as Stereo. On July 23, 2012, the Stereo satellites captured images of one of the most rapid CME events ever recorded with this X-type event spewing mass out from the Sun at a super fast speed of over 2,000 mile/second (3,200 km/second) or 7.2 million miles/hour (11.5 million km/hour.)

NASA and NOAA have set up a system to measure in "real time" the data related to solar wind in a series of "dials" that display the speed, pressure density, plasma heat, and magnetic fields associated with solar wind and CME events. Typically the speed level is in the range of 1–1,000 miles per second, but the July 23, 2012, event was clearly off the charts of the normal velocity data that NOAA's website (using data from the NASA Advanced Composition Explorer, ACE, satellite, launched in 1997) normally displays every 15 min. This ACE satellite, located in the

L-1 Lagrangian point about 1.5 million km (just under a million miles) from Earth is equipped with six sensors and three monitoring devices that allow a wide range of data to be collected and relayed back to Earth so that solar weather and particularly CME events can be characterized in terms of velocity, pressure, magnetic force, and heat. Figure 6.3 shows an artist's representation of a solar storm with Earth being the small blue circle from which the magnetosphere originates.

Today one can contrast and compare the images taken by SOHO and the two Stereo satellites as well as collect the data from the ACE satellite to understand more precisely why and how space weather events occur and also to witness CME (coronal mass ejections) representations in three dimensions, since the Stereo satellite and SOHO shows these events from all angles (Fig. 6.4) [27].

Fig. 6.3 Artist representation of solar wind interacting with Earth's magnetosphere (Graphic courtesy of NASA)

Fig. 6.4 One of the Stereo satellites imaging the Sun (Image courtesy of NASA)

The combined capabilities of SOHO and the stereo satellites have now revealed the workings of the Sun to the greatest level yet attained. The power and the impact of CME events on Earth seem to depend on a variety of functions that include the mass of the solar flare, its rate of acceleration, and its direction in relationship to Earth and Earth's orbital speed.

Despite all of the new knowledge that has been acquired and the greater ability to predict when solar coronal mass ejections will occur there is still a great deal more to be learned. There is likewise a great deal more to be learned about how to protect satellites from these massive solar flares both by designing greater protection and improving operational procedures for when warnings of solar flares are received. The precise nature of how solar storms affect satellites and creates failures is still an evolving field of study. Although satellite operators claim that few total satellite failures are directly related to coronal mass ejections, the statistics show that nearly half of all satellite failures that have occurred in the last twenty years transpired during the years of maximum solar activity in the 11-year cycle seems to be more than a coincidence.

Of greatest concern of all is to understand the threat that coronal mass ejections pose to the safety of people within Earth's atmosphere. Today, airlines that fly polar routes to reach destinations faster and to save fuel, have learned to alter these polar routes when there is a major coronal mass ejection event or gamma ray and solar radiation are at elevated levels.

The issue of elevated radiation and implications of the variation in the ozone layers and the varying sizes of the so-called polar ozone holes are addressed in the following chapter.

Chapter 7
Space Radiation, the Ozone Layer and Other Environmental Concerns

[I]t is our way of just seeing the push and pressure of the cosmos.

–William James

The Problem of Cosmic Ultraviolet Radiation

This chapter continues the discussion that was started in the previous chapter with regard to solar events and activities that provide hazards to people here on Earth. What is clear is that solar and cosmic radiation from other sources beyond actually poses a hazard to flora and fauna on Earth. Most certainly this high-powered radiation provides a particular danger to astronauts. There is continuous radiation reaching Earth from the Sun and cosmic sources throughout the universe. These rays are distorted from a direct hit on Earth by Earth's geomagnetic field.

Powerful and intense ultraviolet and cosmic radiation, particularly high energy gamma rays if not bent away by Earth's geomagnetic, would create a deadly environment on Earth. In addition there are clouds of poison gas in the upper atmosphere such as hydrogen cyanide that are warded off by the geo-magnetosphere as well.

Super energetic X-rays travel from the Sun and across interstellar space. This radiation is continuously raining down from the cosmos and may be elevated during solar max and during CME events, but this is a natural phenomena that is always present. Without the Van Allen belts, Earth's atmosphere, stratosphere and ionosphere, radiation would be a true killer for virtually all life on Earth. When we search for life in outer space we must not only look for Earth-like planets in terms of size, water, and warmth, but planets with a protective atmosphere as well.

The ultraviolet and X-ray radiation from the Sun are a particular hazard to astronauts who are exposed to high levels of damaging rays that those of us who live within Earth's protected atmosphere are not. An astronaut on-board the International Space Station may receive an exposure to radiation that is much higher to someone on Earth and thus radiation levels have to be carefully monitored. In the case of

J. N. Pelton, *Space Debris and Other Threats from Outer Space*,
SpringerBriefs in Space Development, DOI: 10.1007/978-1-4614-6714-4_7,
© Joseph N. Pelton 2013

surges in radiation levels there is a protective compartment on-board the ISS where astronauts and cosmonauts can seek shelter. If at a future date there were to be a permanent lunar colony, it would likely need to be built below the surface to protect the astronauts and cosmonauts from the harmful radiation and high energy mass ejaculates from the Sun characterized as "space weather".

This high energy radiation can also be quite harmful to humans here on Earth. The ozone layer helps protects all flora and fauna against genetic mutations triggered by cosmic radiation in addition to the vital protection provided by the inner and outer Van Allen belts. The screening out of ultraviolet rays by solar backscatter from the naturally occurring ozone layer helps to reduce the ultraviolet radiation threat to humans and other animals in terms of both skin cancer and genetic mutation. If "ozone holes" now significantly detected in the polar regions should extend their diameter to a greater size then humans and indeed most animals would most certainly be at increased risk of both genetic mutation and cancer.

It is through space research that the holes in the protective ozone layer were discovered. NOAA has launched nine such sensors on their meteorological satellites. These sensors are geared to detect the level of backscatter of incoming solar ultraviolet radiation in the ionosphere and provide the amount of backscatter that occurs between the altitudes of 6–30 miles (9.6–48 km). These solar radiation sensors called the SBUV/1 and 2 have been flying on the NOAA satellites dating back to the mid 1980s [28]. This monitoring of the "space weather" from the Sun is key in many ways, including the fact that the solar energy released impacts the weather here on Earth. The 11-year cycle as shown here closely mirrors the electromagnetic flux levels that are released from the Sun as well (Fig. 7.1).

These sensing instruments (shown in Fig. 7.2) have confirmed the existence of the ozone holes in the polar regions since 1987. These sensors confirm that penetration of the UV radiation at these high latitudes are indeed dangerous.

These are not theoretical observations but confirmed facts that carry with them true health and medical implications. Currently residents in the most extreme southern locations of Australia and New Zealanders, for instance, today report a much higher incidence of skin cancer. In my own travels to Adelaide, Australia, in order to teach for the International Space University, it is difficult to venture out during Aussie summer days without a hat and significant amounts of sunscreen in light of the Sun's intense radiation effects. Frogs and amphibians in these polar areas have also demonstrated a growing number of genetic mutations.

Earth is protected not only by the ozone layer that resides above the stratosphere but below the ionosphere and the two Van Allen belts as well. Without the protection afforded by these two high energy belts that are shaped by Earth's magnetosphere and the ozone layer, the plant and animal life of Planet Earth would be subjected to a lethal amount of radiation and plasma that would in a short time likely render our world lifeless. These Van Allen belts are represented in Fig. 7.3 below.

Although the inner radiation belt was discovered a half century ago by the Geiger counter on board the *Explorer I* spacecraft, and the outer belt was discovered a few years later, there is much we still do not know; the make-up and function of these two belts now named after James Van Allen is still not well understood.

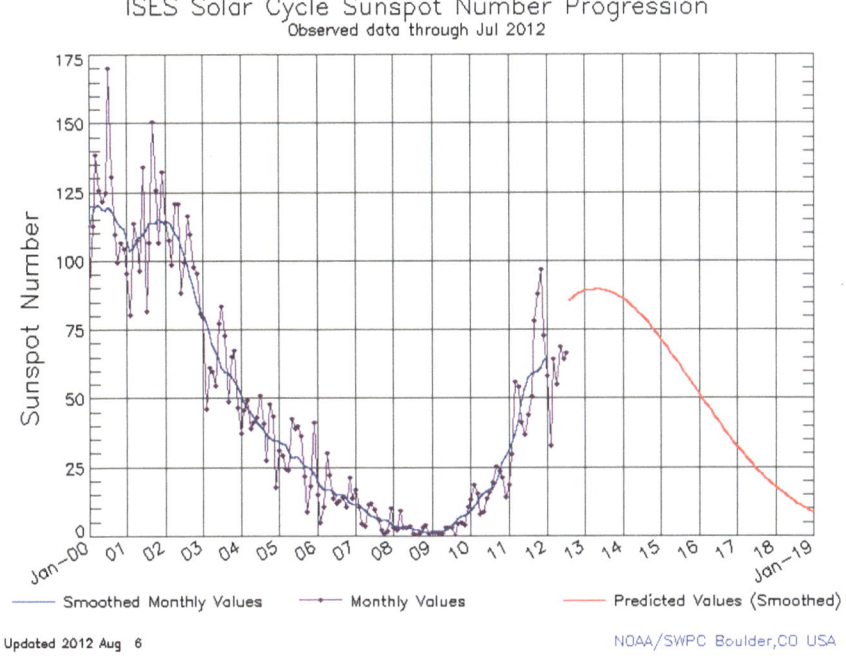

Fig. 7.1 The 11-year solar cycle that also reflects CME activity and radiation flux levels (Graphic courtesy of NOAA)

Fig. 7.2 Solar backscatter ultraviolet radiometer that flies on NOAA weather satellites (Photo courtesy of NOAA)

Two identical space probes known as the Radiation Belt Storm Probes were launched in August 2012 to try to understand in much greater depth the way these belts function and why the belts are so different. The two spacecraft, now renamed the Van Allen solar storm probes, have now detected a third belt that "appeared" and then "disappeared" during the early part of 2013. These research probes will fly virtually identical orbits so that the fluctuating conditions within the belts can

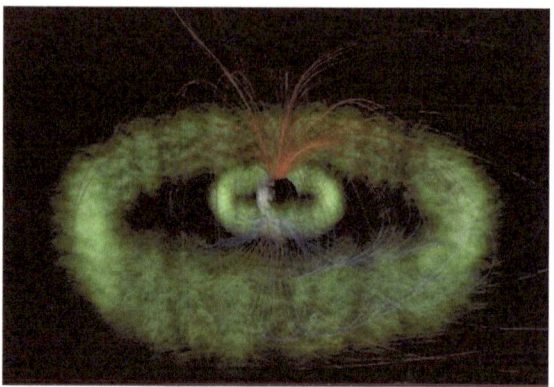

Fig. 7.3 Model of the "invisible" inner and outer Van Allen belts that surround Earth (Graphic courtesy of NASA)

be monitored. It is hoped that the data collected will indicate how the belts interact with Earth's magnetic field as well as how they "absorb" solar and cosmic radiation, capture solar ionic materials from coronal mass ejections and space weather and then dissipate it back into outer space.

The Van Allen belts are in a constant state of flux. The inner belts largely consist of protons at high energy levels while the outer belts largely consist of high energy electrons. These belts grow and shrink as they are affected by space weather. In order to make more detailed measurements, the Radiation Belt Storm Probes will use twin satellites following identical paths, sometimes zipping just above Earth's atmosphere and sometimes many thousands of miles out into space where the outer belt resides. One of these Radiation Belt Storm Probe satellites is seen in the foreground and another one in the far distance is pictured below in Fig. 7.3 as they would appear as seen from above in orbit.

Traveling through both the inner and outer radiation belts, the crafts will maintain varying distances, sometimes within 100 miles of one another, at other times nearly 25,000 miles apart.

Their separation will allow scientists to get a better feel for changes that occur in the belts. One craft may absorb a spike in radiation levels. What the second craft reads from a different location will reveal a great deal about what is happening in the belts.

In some cases it may read a similar increase. It may only detect the higher levels when it reaches the same region the first craft was traveling through. It may pick up the spike after a slight delay, indicating that the radiation is traveling, wavelike, through the belts. Or it may see nothing at all (Fig. 7.4).

Distortions in Earth's Magnetic Field

There are at times magnetic irregularities and distortions in Earth's magnetosphere that normally flows upward from Earth's two magnetic poles that are near the world's geographic poles as well. At times there are fluctuations in Earth's

Fig. 7.4 The two radiation belt storm probe satellites as they would appear in orbit (Graphic courtesy of NASA)

magnetic flux system that can serve to bring in deadly cosmic and solar radiation down to the world's surface.

As early as 1961 James Dungey of the United Kingdom predicted that "cracks" might form in the magnetic shield when the solar wind contained a magnetic field that was oriented in the opposite direction to a portion of Earth's field. In these regions with the two competing magnetic fields can sometimes interconnect through a process known as "magnetic reconnection". This process can serve to form a modest crack in Earth's shield. In this case electrically charged particles of the solar wind as well as ions from the Van Allen belt could flow below the geomagnetic field. This can bring not only deadly radiation but poisonous gases such as hydrogen cyanide.

These small "cracks" were first detected using the International Sun-Earth Explorer (ISEE) satellite as early as 1979. This potentially very serious threat has thus been under study since that time [29].

A joint satellite mission funded by NASA and the European Space Agency, named IMAGE, has been launched to monitor these "cracks" and to determine the degree to which Earth's geomagnetic field might be weakening and thus increasing these dangerous conditions over time. The IMAGE satellite, as pictured in orbit inside of one of these magnetic gaps, is depicted in Fig. 7.5.

If indeed Earth's geomagnetic field is weakening or if the polarity of the field is switching north to south and vice versa, then it is possible that all of humanity and

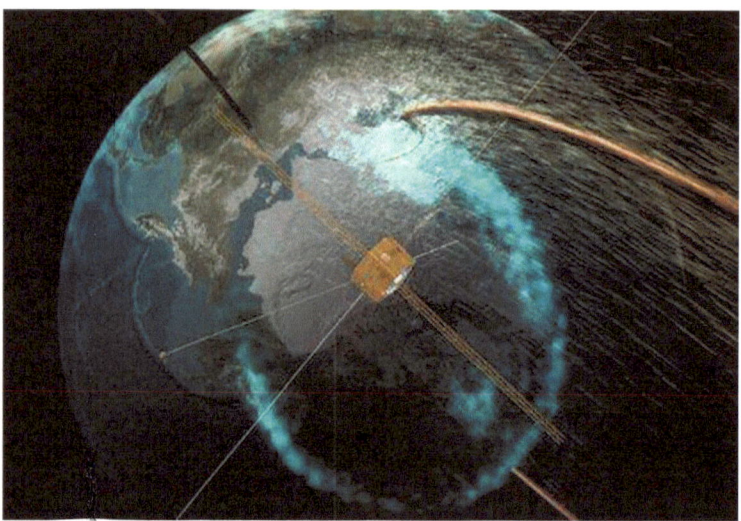

Fig. 7.5   Artist representation of the IMAGE satellite in orbit (Graphic courtesy of NASA)

indeed plant and animal life might be increasingly exposed to energetic electrons, ionic particles as well as deadly gases from the inner Van Allen regions [29].

In the very first days of 2011 there were a number of events around the world—Arkansas and New Orleans in the United States, Italy, New Zealand, Japan, etc., where it seemed that perhaps millions of birds died in a "freak event". This event is still not fully explained, but one persistent theory developed by Russian scientists is that a deadly cloud of hydrogen cyanide gas was leaked through a crack in Earth's geo-magnetosphere at an altitude of some 50 miles high. The theory is that enough of this hydrogen cyanide gas leaked through to a lower altitude and this was the cause of the mass death of so many birds all at once. At the same time there were a number of deaths of fish as well. The theory is that the clouds of deadly gas managed to rain down this hydrogen cyanide poison as rainwater and that this mixture was still sufficient to kill freshwater fish as well [30].

Chapter 8
NEOS, Comets, Asteroids, Bolides and Meteors

...but in reality there are only atoms and space.

–Democritus of ancient Greece

The official definition of a "mass extinction event" is one that wipes out over 30 % of the species on Planet Earth. Of the five mass extinction events that have occurred over the last half billion years, the last one is believed to have been caused by a massive asteroid, estimated to have been some 6 miles (10 km) in diameter. It hit the ground with a force equivalent to over 6 billion Hiroshima-sized atomic bombs, or one such bomb for every human on Earth. This massive asteroid created a crater some 62 miles (100 km) in diameter, as shown in Fig. 8.1. It created searing heat equivalent to the surface of the Sun and instantly killed by heat and concussive shock everything for a huge distance around. But this was just the first phase. It created hurricanes with winds of 500 miles (800 km) an hour and raging forest fires.

The ultimate killer that managed to wipe out up to 70 % of the plant and animal species on the planet was the deadly cloud of iridium-rich debris that screened sunlight from the surface for many months. This absence of sunlight was the largest killer of all. The strike that was visible for only one second before impact not only ended the age of dinosaurs but also killed off almost everything else. The crater from such a massive hit to Mexico and the Caribbean is still visible today. This is a visible reminder that it can happen here. Asteroid and comet strikes are not just a concept for a science fiction screenplay but a very threatening reality [31].

If a K-T event were to happen today, the destruction to buildings, farmlands, dams, energy, information networks, and more would be vast. Perhaps most of humanity would be quickly returned to the Stone Age. Many plants, wildlife, livestock, and people would die due to the initial heat and energy, but the loss of sunlight would spread death for perhaps the majority of species across the planet. The losses in economic terms would not be in the trillions of dollars but the quadrillions of dollars. As death spread across the world, money would become

J. N. Pelton, *Space Debris and Other Threats from Outer Space*, SpringerBriefs in Space Development, DOI: 10.1007/978-1-4614-6714-4_8, © Joseph N. Pelton 2013

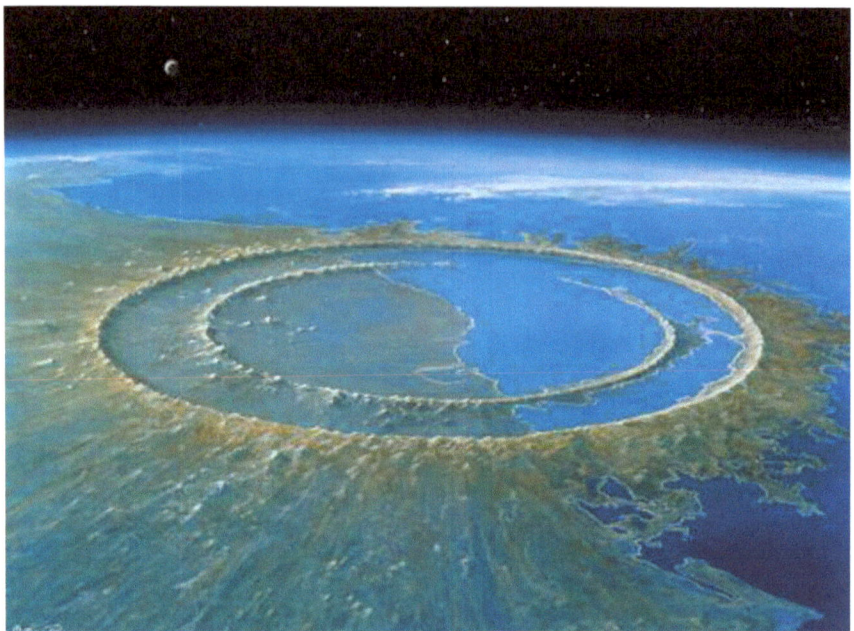

Fig. 8.1   The crater from the killer asteroid that wiped out 70 % of the species on Earth

meaningless as survival began to trump the concept of personal wealth. A repeat of the K-T event would be a very, very, very bad thing indeed. It is thus imperative for astronomical and space programs to do everything possible to detect such threats and to take preventive action against such a dire event.

At this time some 90 % of the asteroids that are greater than 1,000 m in size are thought to have been identified, but only 20 % of the asteroids in the range of 100–1,000 m have been located. A space object of this size is still quite massive. An asteroid, such as a known near Earth object named Apophis measures only about 300 meters in size, a small fraction of the asteroid that triggered the K-T extinction. Yet if Apophis were to hit Earth at 40,000 miles per hour (64,000 km/hour) would still do very substantial harm. NASA's calculations indicate that it would have the equivalent power of 30,000 atomic bombs exploding. If this giant bolide were to hit the Atlantic Ocean near the East Coast, the ensuing tsunami would literally wipe out Boston, New York City, Washington, D.C., Miami and indeed every other coastal city, town and village in between. But this would only be the initial fatalities. Potentially billions of people could be lost from a direct hit from an Apophis-sized asteroid and the cloud of dust that would enshroud the planet for months to follow. Recently a meteorite the size of an aircraft carrier with an estimate size of 10,000 tons burst above Siberia, Russia with tremendous force. Thousands of people were injured and tens of thousands vehicles, homes and buildings damaged. This of course was like a pebble in relation to potentially hazardous asteroids are out there, but this was a wake up call to those who think this danger is imaginary.

Although the man in the street does not think about the threat from asteroids and comets, space scientists at NASA, ESA, and JAXA certainly do. The WISE space telescope launched by NASA in mid-December 2009 has been carefully mapping as many NEOs as possible during its time in orbit as well as far flung galaxies as well. In the time between its launch and its shut down on February 17,2011, the Wide-field Infrared Space Explorer had collected millions of images and completed one and a half inventories of Earth's overhead sky by a systematic scanning process. The WISE satellite infrared sensors during its mission captured some 1.8 million images. These have allowed scientists to detect nineteen previously unseen comets. It also allowed the detection of over 33,500 asteroids and 120 previously unknown near-Earth objects (NEOs) that could become potential hazards to Earth at a future date. The infrared sensors that detect radiation outside the visible light range was able to detect low heat dwarf "brown stars" and detect objects that might be invisible due to dense dust cloud layers and other obscuring elements.

The quarter-of-a-billion-dollar project was certainly successful in living beyond its projected 10-month mission lifetime. Ultimately it was the exhaustion of the coolant for the infrared sensors that was the lifetime limiting factor. By cleverly shifting from a four IR sensor operation to only two the lifetime was extended further than expected. After the coolant was entirely expended, a further program called NEO-WISE was undertaken for three months up until its February 2011 end date. During this NEO-wise phase, the spacecraft was entirely devoted to searching for NEOs. But this task still remains to be completed.

Currently this comet- and asteroid-detecting spacecraft is now in hibernation with its coolant expended. In order to complete this crucial inventory of potentially dangerous asteroids and comets that might eventually hit Earth, additional space telescopes are needed. This would likely mean satellites with the ability to alter their range of view, more IR sensors and sufficient coolants to extend the spacecraft's lifetime. The question that many would ask is: "Is it really necessary to spend this much money on the very long shot that we might find a killer asteroid?" It turn out there is a fairly good answer to that question.

Fig. 8.2 The potential future dangerous path of 2011 AG5

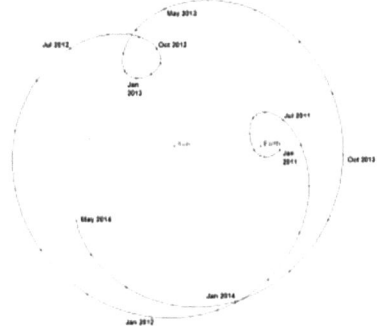

Asteroid 2011 AG5 was discovered in January 2011 by the WISE imaging process. After initial analysis it was determined that there was a very credible chance that Asteroid 2011 AG5 could indeed collide with Earth in 2040. After further analysis it was decided that this was a long shot indeed unless something very strange happens in terms of the asteroid's interaction with the Sun's gravity, known as a "keyhole" event. This "keyhole" gravitation event that would make the asteroid do a "loop-de-loop" in a way that could put this asteroid in a resonance orbit with Earth. If this should happen it could result in about a 15 % chance of a horrific collision in 2040. The problem at this time is that this PHA (potentially hazardous asteroid) is on the other side of the Sun, and thus no precise measurements can be taken until late in 2013 or 2014. Figure 8.2 indicates the possible "keyhole" event and how this could actually spell trouble down the road.

Asteroid Diversion

If it turns out the 2011 AG5 goes through the gravitational keyhole in just the wrong way, and it is set to hit Earth like multiple atomic bombs, what could we do about it? Well, the European Union has launched an admirable new multi-national research project involving efforts to develop better ways to divert the course of "killer asteroids".

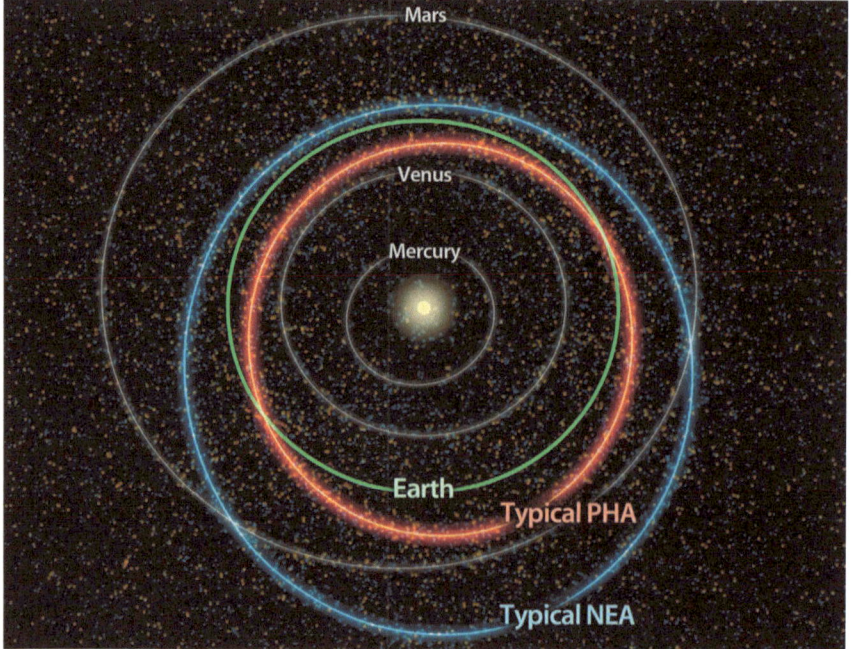

Fig. 8.3 The orbits of NEOs and PHAs

THE TORINO IMPACT HAZARD SCALE

Assessing Asteroid and Comet Impact Hazard Predictions in the 21st Century

No Hazard (White Zone)	0	The likelihood of a collision is zero, or is so low as to be effectively zero. Also applies to small objects such as meteors and bodies that burn up in the atmosphere as well as infrequent meteorite falls that rarely cause damage.
Normal (Green Zone)	1	A routine discovery in which a pass near Earth is predicted that poses no unusual level of danger. Current calculations show that the chance of collision is extremely unlikely, with no cause for public attention or public concern. New telescopic observations very likely will lead to re-assignment to Level 0.
Meriting Attention by Astronomers (Yellow Zone)	2	A discovery, which may become routine with expanded searches, of an object making a somewhat close but not highly unusual pass near Earth. While meriting attention by astronomers, there is no cause for public attention or public concern, as an actual collision is very unlikely. New telescopic observations very likely will lead to re-assignment to Level 0.
	3	A close encounter, meriting attention by astronomers. Current calculations give a 1% or greater chance of collision capable of localized destruction. Most likely, new telescopic observations will lead to re-assignment to Level 0. Attention by public and by public officials is merited if the encounter is less than a decade away.
	4	A close encounter, meriting attention by astronomers. Current calculations give a 1% or greater chance of collision capable of regional devastation. Most likely, new telescopic observations will lead to re-assignment to Level 0. Attention by public and by public officials is merited if the encounter is less than a decade away.

Threatening **(Orange** **Zone)**	5	A close encounter posing a serious but still uncertain threat of regional devastation. Critical attention by astronomers is needed to determine conclusively whether or not a collision will occur. If the encounter is less than a decade away, governmental contingency planning may be warranted.
	6	A close encounter by a large object posing a serious but still uncertain threat of a global catastrophe. Critical attention by astronomers is needed to determine conclusively whether or not a collision will occur. If the encounter is less than three decades away, governmental contingency planning may be warranted.
	7	A very close encounter by a large object, which if occurring this century, poses an unprecedented but still uncertain threat of a global catastrophe. For such a threat in this century, international contingency planning is warranted, especially to determine quickly and conclusively whether or not a collision will occur.
Certain **Collisions** **(Red Zone)**	8	A collision is certain, capable of causing localized destruction for an impact over land or possibly a tsunami if close offshore. Such events occur on average between once per 50 years and once per several 1,000 years.
	9	A collision is certain, capable of causing unprecedented regional devastation for a land impact or the threat of a major tsunami for an ocean impact. Such events occur on average between once per 10,000 years and once per 100,000 years.
	10	A collision is certain, capable of causing a global climatic catastrophe that may threaten the future of civilization as we know it, whether impacting land or ocean. Such events occur on average once per 100,000 years, or less often.

Fig. 8.4 Torino impact scale for potentially hazardous asteroids

The three prime areas of research are exploring the use of gravitational attraction for course diversion, "bombing" the asteroid out of existence, or hitting it with a missile. This program is called NEOShield. The problem is that the four-year program is funded at a very inadequately low level of 4 million Euros. We spend billions on national defense and medical research against pandemics. The funding for NEOShield is a mere pittance. We need to be spending at least 10 times more to produce any real hope of viable results.

There are dozens of meteor showers that occur each year as Earth orbits the Sun, since there are literally millions of small meteors in solar orbit. Larger meteoroids are called bolides and even larger ones still are called asteroids. It is the larger scale near-Earth asteroids (NEAs) and particularly the PHAs that come within 5 million miles (8 million km) that are the ones that are of the largest concern. Typical orbits for these asteroids and how they could intersect Earth orbit are shown in Fig. 8.3.

Since it has been 65 million years since the K-T mass extinction event there is some reassurance that another such horrendous event is a very remote possibility. Nevertheless it is estimated that some 10 % of the really big PHA's have still be discovered, and 80% of the PHSs in the 100–1,000 m range have not been identified. These projections are based on the NEO-WISE program activity and calculations undertaken based on the NEO-WISE program activity and calculations undertaken based on its observation of just one sector of the sky. Space scientists have take the potential threat seriously enough to adopt a hazard scale as provided in Fig. 8.4 [32].

This scale seeks to set the risk of potential collisions between asteroids and Earth in some perspective and indicates that while collision events that are in the range of 1–3 occur fairly frequently, the impact on the overall world is small. It is the frequency likelihood for events in the 9–10 range that is problematic in placing the range between 10,000 and 100,000 years [33]. The K-T event, which would be like a 12 on this Torino Impact Scale took place, of course, 65 million years ago.

The logical question that springs to mind with regard to these asteroid threats—or for that matter comet collision threats—is what can be done about them? Can we do more than just identify what are potentially the most hazardous objects out there and when could they potentially collide with Earth? The various strategies for addressing these threats from outer space are addressed in the following chapter.

Chapter 9
Protecting Against Comets and Potentially Hazardous Asteroids

From a wild weird clime…sublime, Out of Space—out of Time.
–Edgar Allen Poe (1845)

The Odds are with Us: Unless We Miscalculate

We know that there are on the order of 40,000 so-called NEOs orbiting the Sun. At least these are the best calculations that come from a reasonable sample of the sky based on the observations and thousands of pictures taken by the WISE spacecraft.

And these NEOs are just the ones that are 100 m or larger in size and have orbits that come within some 9 million miles (14.4 million km) of Earth. The ones that are truly potentially hazardous are fortunately significantly less in number.

Even with all the data collected by Earth observations and scientific spacecraft like the WISE vehicle, there is still a great deal that we do not know. In short, predicting those asteroids or comets that truly might collide with Earth is not as easy as it might seem. There are distortions in the Sun's gravity that might make a much smaller and less massive object go through a so-called "keyhole" event such as portrayed in Fig. 8.2. Also, a PHA that comes reasonably close to Earth on a particular orbit can clearly be disturbed by Earth's own gravity so that when it returns some years later the impact becomes more likely. This is the case of Apophis, which will come past Earth in 2029 in such as way that it could come closer in 2036.

Although this is even more unlikely it is also possible that the gravity from the Moon or even Venus or Mars could create a disturbance that results in a later collision with Earth many years later. There is further uncertainty about an asteroid's path as a result of the so-called Yarkovsky effect. This effect is named after the Russian engineer who first identified in 2003 how the Sun affects an asteroid's orbit other than just by gravitational pull. As an asteroid travels around the Sun it is heated by the solar rays. After a time the asteroid begins radiating energy away

J. N. Pelton, *Space Debris and Other Threats from Outer Space,*
SpringerBriefs in Space Development, DOI: 10.1007/978-1-4614-6714-4_9,
© Joseph N. Pelton 2013

from itself in such as way that it very slightly alters its flight path. Although this is a very subtle phenomena it adds up over many years to become a real feature in orbital path prediction [34].

The NASA Safeguard Program and the European Space Agency NEO Shield Program are researching and developing strategies for protecting Earth against a future PHA or comet strike. Some of this research is geared to finding out what materials constitute the asteroid. This could be metallic, granite or stony or some other material that might be less massive and easier to divert. Another part of the research is geared to the shape of the asteroid and to determine if it might be more easily split or diverted by one or more means. In short some strategies for diversion of the orbit of an asteroid or comet depend a good deal on both the shape and the composition of the object in question.

Although there are a number of strategies that might be employed the top four strategies involve the following:

1. *Crashing a rocket with a heavy payload into the PHA or comet at a very high relative velocity.* This strategy is the simplest and likely the most cost effective one that might be employed. It would slightly alter the mass and gravitational pull on the threatening body so that over time the orbital trajectory would change. Of course there could be variations on the theme, such as to crash two rockets into the offending space object to double the gravitational effect. In order for this strategy to be successful precise navigation would be required to ensure that contact would be made and with sufficient impact so that the mass of the rocket and its payload would remain in contact over a period of months or even years so the orbital path would change.

2. *Tethering a rocket to a PHA or comet with sufficient chemical fuel or electronic propulsion to alter its orbit.* This strategy is actually a variation on the theme of the previous approach. In this case the harpooned space object would not only supply additional mass that would serve to change the asteroid or comet's orbit, but the remaining chemical fuel could be expended to either push or pull the body into a less threatening orbit. In addition a long acting electronic ion engine could also add some propulsion to steadily move the body toward a safer orbit. The electronic thruster might also be used to maneuver the rocket into place for the harpoon firing to establish the link with the rocket. This approach would require even more precision guidance than the simpler approach of crashing the rocket into the threatening object, but the effect could be more substantial over the longer term.

3. *Use of a nuclear weapon to divert the asteroid or comet.* This strategy would require much more knowledge of the asteroid or comet. In short one would need to know about the shape and composition of the object. Hopefully one would be able to place with some accuracy a shaped explosive or tactical weapon onto the threatening NEO in such a way that the nuclear device would generate a predictable new orbit and avoid simply creating several fragments that could also endanger Earth. The strategy could thus be quite crude and unpredictable in terms of a desperation action, or it could be a much

more well-engineered plan. The more sophisticated approach might start with a smaller chemical charge first being detonated by robots at a predetermined location. This would be followed by a tactical nuclear device being put in place to create a second much more powerful shaped explosion that would initiate a pre-planned new orbit. This could be carried out as a mission with human crew aboard or perhaps conducted entirely with robots. In a sufficiently dire situation, there might be volunteers who would be willing to sacrifice their lives on a one-way mission.

4. *Use of reflective devices to vaporize the PHA or comet's surface so as to create new thrust and shift the orbit.* The feasibility of this approach clearly needs to be tested, and certainly there is much that could go wrong in terms of deployment of metallic reflectors, the orientation of the reflectors so that the solar rays are properly oriented to the surface of the PHA or comet, and even the degree to which the ablated surface would create sufficient thrust to achieve the desired orbital changes. There is a new variation on this approach that would create a phased array of high powered lasers with enough strength to vaporize the asteroid from an Earth based location. This system known as DE-STAR for "directed energy solar targeting of Asteroids and exploRation" would use large scale solar reflectors to bombard and vaporize asteroids. This same system could also be used at the same time to generate energy that could power vehicles to travel within the solar system.

The wisdom of Occam's razor seems especially compelling in the field of space in advising that if there are several solutions then pick the simplest.

In all of the above possible solutions clearly time and forewarning is of the essence. The earlier the potential threat is identified the easier the diversion of the orbit is to achieve. If there can be years of early warning, then the very slightest change in orbital vectors can allow safe passage away from Earth. Conversely, if there should be just a very short period of warning, then the diversion becomes incredibly more difficult. Future satellite surveyors such as those similar in design to the WISE spacecraft can certainly aid in identifying possible threats to be cataloged and studied.

Today the greatest danger appears to come from the influences that can throw off orbital projections and perhaps convert a seemingly harmless NEO into a massive killer and possibly catching us off guard with too little time to respond. The so-called "keyhole gravitational affect" of the Sun on the massive 2011 AG5 or possibly the 1999 RQ36 asteroids are clearly potential "jokers in the deck" that could result in our not being warned in sufficient time. This "keyhole" gravitational phenomenon is to be closely watched and hopefully better understood. Yarkovsky effect also needs to be better understood. Finally we need to have a better sense of what techniques, under what set of circumstances, make the most sense to use on threatening NEOs.

Chapter 10
Top Ten Things to Know About Threats from Outer Space

In centuries when our population was smaller, our species could blithely pollute the air and water of the Earth and remain unscathed by the consequences. So too, for the last several decades we thought we left our space debris in orbits unlikely to collide. We are only slowly realizing the error of our ways.

–Paul S. Dempsey, Director, Institute of Air and Space Law,
McGill University

Coping Strategies for Space Threats in the Near and Longer Term

Large and immanent threats that could result in large-scale death and destruction tend to dominate the news. Coming mega-disasters vividly capture the public attention because people and their loved ones are at clearly risk, and there is preciously little time to prepare for the worst.

These pending disaster events often tend to command major budget allocations either for prevention or recovery—perhaps both. This is the case whether one is talking about a volcanic eruption, a Category 5 hurricane just off shore, or even rapidly rising unemployment in an economy that is in a financial nosedive.

Other types of threats are more difficult to sort out and prepare for intelligently. Clearly minor threats that are very far away get virtually no attention at all. Medium term threats obviously rise in public awareness as they come closer to happening, and this is particularly so if the threat is clearly manifest and easy to understand. Forecasts of particularly high tides for a hotel on the Grand Canal in Venice, Italy, is taken most seriously as the day of the event draws near because the coming flood damage is quite real, and complete recovery may take weeks to accomplish before the hotel can reopen.

J. N. Pelton, *Space Debris and Other Threats from Outer Space*,
SpringerBriefs in Space Development, DOI: 10.1007/978-1-4614-6714-4_10,
© Joseph N. Pelton 2013

Perhaps the most difficult-of-all threat to cope with is the one that is hard to understand and visualize, seen to be far away, and yet if it occurs could be truly devastating to the future of humanity and its ultimate advancement. Yet such threats indeed do exist. It is possible that the buildup of orbital debris over time could threaten all future science and application satellite deployments. Runaway cascades of orbital debris could in time make it impossible to launch astronauts safely or to launch defense satellites to sustain national defenses.

Beyond space debris there are other "obscure threats" from space that could wipe out our modern electronic grids, disable millions of electronic processing devices such as those in cars, trucks, airplanes, and home and business computers. These are events such as super solar flares (or coronal mass ejections), a surge of solar or cosmic radiation, or cosmic events that could disturb Earth's magnetosphere in a significant way. Finally the most unlikely (in a statistical sense) cosmic event, yet potentially the most devastating of all, is the risk of a huge NEO such as a comet or a potentially hazardous asteroid crashing into Earth with the force of millions of atomic bombs and wiping out human civilization and most animal and plant life along with it.

This book has attempted to acquaint the reader with the nature of these threats, the likelihood of their occurrence, the various strategies that might be undertaken to cope with these threats, and best steps forward. The following are some of the more important take away lessons from this book.

1. We are far from solving the problem of orbital space debris.

Figure 2.1 shows just how dramatic the increase in orbital debris has become. Although tons of debris from LEO decays each year, the mass of satellites launched into Earth orbit each year far outweighs the de-orbiting debris. Although there have been reforms to eliminate elements of debris such as exploding bolts and such, there are still upper stage rocket motors, nose-cone fairings that cover satellites during ascent, etc., that are left orbiting as debris in space. Each launch, even with voluntary guidelines to control new debris, still creates perhaps 10–12 new debris elements. Satellites in LEO and Sun-synchronous polar orbits, where the greatest problems lie, are generally equipped so as to be actively de-orbited or launched to an altitude where relatively rapid orbital decay will naturally occur.

But these de-orbit plans are not perfect. Problems can and do arise. In the case of the Iridium satellites for instance there were several problems. In some cases it was not possible to command a satellite to fire the jets to de-orbit. In one case Iridium officials indicated that they were asked not to de-orbit a satellite because of its risk of colliding with a U.S. defense department surveillance satellite.

If there are 10 new debris elements created with each launch and over 100 launches a year this could generate 1,000 new debris elements or 10,000 over a decade. But the debris accumulation occurs in many other ways. One collision between two large space objects or a satellite and a missile can generate 3,000 new debris elements. A fuel tank explosion can also generate hundreds of new debris elements.

All of these problems obviously mount up. Missiles hitting satellites, satellite or large rocket engine collisions, and fuel tanks exploding are the biggest source of problems. Experts have indicated that with no new launches the total number of debris elements will significantly increase over the next decade without active removal processes.

This is not a problem that is going to fix itself. The long-term sustainability of space is indeed at risk due to mounting orbital space debris. Although the prime concern is in the LEOs and particularly the Sun- synchronous polar orbits so key to meteorological and remote sensing satellites, there is a concern with mounting orbital debris in all orbits. The concern has increased to the degree that application satellite operators have formed the Space Data Association (SDA) to seek to avoid the collision of active satellites. The United Nations Committee on the Peaceful Uses of Outer Space (COPUOS) has also formed a working group on the Sustainability of Space to seek solutions to orbital debris problems and other issues that might impair the future uses of space for all nations. Today only some ten to twelve countries can launch spacecraft into orbit, but all nations benefit from spacecraft and their uses for communications, navigation, remote sensing, meteorology, geodetics, scientific research, and national defense.

2. There is no internationally agreed mandatory system to prevent the buildup of space debris nor easy to administer international agreement under which removal of space debris can be easily achieved.

Two international groups have now adopted voluntary guidelines to reduce the formation of new orbital debris. These groups are COPUOS and the Inter-Agency Space Debris Coordination Committee (IADC). The IADC is an international governmental forum of space agencies that are working together to mitigate and minimize the adverse affects of manmade and natural debris in space. Unfortunately, because of the complexity of international coordination it has taken 18 years to develop and get groups to agree to voluntary procedures to reduce debris. These voluntary procedures do not address the issue of nosecone fairings or upper stage rockets.

There is currently no obligation for those who launch satellites to ensure that space debris is removed, and the liability treaty actually represents a barrier to countries having others undertake to remove any such debris, since liability cannot be transferred.

There have been proposals that countries require launching organizations sign an "assured removal" clause as a condition for obtaining a license to launch and operate a satellite regardless of whether they were deployed by a national or foreign launch service. In the case of GEO debris removal this could mean placing the satellite in a "graveyard" orbit. Such a clause would apply not only to the satellite but the upper stage of the launcher and its fairings and also require that insurance be obtained if for some reason removal was not accomplished.

Alternatively it has been proposed that all launching organizations pay into a fund dedicated to orbital debris removal or mitigation of new debris being formed. For either one of these proposals to become effective a change in the international space liability treaty is necessary.

3. There is no clearly identified and tested technical/operational means by which orbital debris can removed from orbit.
Further, any such means would be: (i) expensive to carry out; (ii) difficult to accomplish and administer under existing international law and (iii) lacking any attractive way to finance such a removal process, which will be expensive to execute under virtually any plausible scenario.

The change to international space liability convention is one of the steps needed to start cleaning up the various orbits in space. But the even larger hurdle appears to be the challenge of developing the technology that can allow the cost-effective and reliable removal of space debris from orbit, especially from LEO and Sun-synchronous polar orbits. The great variety of concepts that might be used for this purpose was outlined earlier in the book. Unfortunately none of these technologies or concepts are considered to be mature and ready for near-term implementation.

Any technologies that require a separate launch for each major debris element removed are obviously quite expensive. There are variations on the concept of electro dynamic-based systems that would utilize Earth's magnetic field to derive electric propulsion to power such removal operations that might prove to be more cost effective. The Slingsat concept that would use momentum generated from the removal process might also prove to be less costly.

For any of these concepts to ultimately prove viable would require for the method to be broadly endorsed by the international community to remove a large number of satellites. To date no such broad agreement exists.

On top of all of the technical challenges, there is yet the further challenge that many of the space-based debris removal systems or the ground-based laser pulses could be considered to be "space weapons". One of the innovative solutions suggested in this respect is to have the country of launch registration and "ownership" of the space debris assume direct operational control of the removal or avoidance process so that they directly oversee the removal activity.

4. There might be some new and innovative ways to address problems associated with orbital space debris that include using ground based lasers to avoid collisions of larger space objects and financing removal operations.
Another approach that has been suggested as an interim solution, until effective removal processes are developed and tested, is to simply direct ground-based laser pulses to satellites or other large space objects on likely collision courses so as to avoid such collisions. Since these objects are moving at very considerable speeds, such as on the order of 25,000 km or (15,000 miles) per hour, relatively small pulses could slow the velocity of the satellite or space object a very miniscule amount to avoid the collision. The problem is sufficiently accurate tracking with laser-based systems to make sure that the pulses would indeed avoid the collision rather than actually causing the unwanted impact to occur.

A variation on this theme of active collision avoidance is already being conducted with regard to the International Space Station and other high value satellites such as surveillance or "spy" satellites. In this case on-board thrusters or, in the case of the

ISS propulsion engines, can be fired to lift the space object above the orbit of the satellite or inert space object that is considered an impact threat. Experiments to test these concepts in terms of the accuracy of laser-based precision tracking and minute adjustments to space objects orbital speeds are now being conducted. The issue of whether such laser bursts might be considered a space weapon, however, remains an open issue.

5. Solar flares and coronal mass ejections (CMEs) could be much larger problems than previously thought with modern electronic and electrical systems particularly at risk.
The documented evidence of the so-called "Carrington event" remains very much a reminder that very powerful solar flares or coronal mass ejections can zap Earth's thin protective atmosphere shield with enormous force. When such a torrent of ions or high energy radiation head directly for Earth at great speed, the Van Allen belts serve to shift the impact toward the polar regions. The atmosphere that protects our planet can still be distorted and extended some ten to twenty times Earth's diameter. In the case of the exceptionally powerful Carrington event this distortion might be as much as 30 Earth diameters and the remaining level of protection becomes very thin indeed until the atmosphere is pulled back by Earth's gravity.

No such truly mega-event has socked Earth since 1859 although the March 1989 event was quite severe. The bottom line is that we have no way of knowing with any precision when the next such massive coronal mass ejection will occur and where this stream of super-charged ions would impact Earth's atmosphere with the maximum velocity and destructive power. Fortunately, less massive CMEs do occur to provide useful data. During the solar max year in the 11-year cycle, particular experience is gained in how to power down satellites when the most powerful coronal mass ejections do occur. Despite some degree of radiation hardening, heavy duty switches and the powering down of satellites when alerts are received of the most violent space weather, failures of satellites still occur.

Here on Planet Earth dangers from the most severe CMEs also exist. Certainly power transformers have been disabled or destroyed by solar events within recent times, and concerns have increased that all sorts of computer processors and electrical systems in space and on the ground could be destroyed on a massive scale if something like the Carrington event were to occur today.

Even more effective protective strategies can be undertaken for space systems. Radiation hard electronic components and wiring, highly protective insulation coating, wide gap on/off switches and circuit breakers, component redundancy and even Faraday cage structures are all part of the strategy to protect spacecraft from coronal mass ejections as well as an electro-magnetic pulse (EMP) from an in-orbit nuclear explosion.

Satellites, perhaps most importantly, can be powered down when there are alerts of a powerful solar ejection. On the ground transformers can be built within Faraday cages and in critical situations could be powered down and again equipped with surge protectors and heavy duty circuit breakers.

The most difficult area of all is with regard to the highly distributed millions of digital processors in automobiles, aircraft, appliances, etc., all over the world that could be zapped by a massive solar storm.

To cope with this level of threat would require a whole new mindset in design all over the world. This would involve changing basic designs so that digital processors would be backed up by mechanical systems that would in case of emergency back up the billions of processors and electrical systems that now permeate our daily lives. This is really not likely to happen unless there was a massive disaster in which billions of vehicles, aircraft and devices across the world were to be suddenly disabled. Short of this there could be underground parking garages and protective hangars and bunkers where critical aircraft and vehicles for national defense and other strategic purposes might be stored.

There was in the past perhaps only a day to day and half warning of such a pending disaster, since such a solar event would be seen within eight minutes of its occurrence. The highly charged particles even traveling at 4 million miles/hour (6.4 million km/hour) would take about 22 h to reach Earth. With the new information captured by the ACE, SOHO, and Stereo satellites, however, it is now possible to develop predictive information that can lead to as much as a three-day warning of a truly major CME event.

The problem that still remains is to convince the public and especially key governments to prepare for such a super solar storm—one of enormous intensity that may occur only once in hundreds of years. If virtually all of the world's electrical grids, automobiles, aircraft, computers and electrical devices were to suddenly shut down, the world would in a matter of hours be virtually reduced back to the Stone Age.

There are several logical steps to be taken. First a concerted effort should be undertaken to ensure that Earth-based global solar monitoring systems and satellites serving this same purpose are sufficiently connected to global alert systems. Second a process should be begun to explore whether there are enough "solar flare-proof" buildings or underground facilities to at least protect critical aircraft, vehicles, telecommunications switches, or electrical grid transformers. In the case of above ground electrical transformers Faraday cages might systematically be built as protective structures. Third national legislatures might move to set safety standards for protection of electrical grids, telecommunications networks and other critical infrastructure.

In this discussion of legislatively mandated standards (or perhaps commercially developed safety protection standards) there might be additional provisions for heavy duty circuit breakers. There could be emergency alert systems that could allow powering down of electrical systems in the case of a super solar flare. It is not likely that cars or aircraft will in the future be built without processors, but perhaps there could be standards offered that allow for their better protection. Consumers might at least be offered options for automobiles with higher levels of "rad and ion hard" protection, and aircraft manufacturers could offer increased levels of protection as well.

6. We need a much better understanding of the hazards of radiation, particularly in an era with a diminishing protective ozone layer.
The greatest danger to spacecraft as well as to ground-based electrical grids, telecommunications networks and even pipeline lines clearly is posed by high energy

coronal mass ejections or possibly even an orbital nuclear explosion that would give rise to an electro-magnetic pulse, but there are other serious threats from solar and cosmic sources. Super-charged X-ray gamma rays that come from the Sun and cosmic sources pack quite a wallop. The Van Allen belts and the ozone layer serve to screen out all but the most energetic of these radiations. The ozone holes in the polar regions allows the ultraviolet rays to penetrate through and constitute a threat in terms of not only elevated cancer risk but also greater risk of genetic mutation.

Gamma and high energy X-rays have been correlated to elevated skin cancer levels in humans in high latitude countries and to mutations in amphibians. These dangers are currently contained to areas with relatively low levels of habitation, but there are concerns that if the ozone holes should continue to expand that these risks could spread to an ever-larger geographic area. Although cancer is clearly a major danger, the spread of genetic mutations to humans over an expanded area could be a substantial threat to the entire human species.

7. Potentially hazardous NEOs pose a creditable threat. A large enough body colliding with Earth could result in a mass extinction of humankind and much of the animal and plant life on planet Earth.
Sixty-five million years is a long time for a species that has existed only a few million years. The so-called K-T mass-extinction event eliminated 65–70 % of the species on Planet Earth, and we know that if the planet today were to be hit by an asteroid of comparable size that it really would knock Earth back to the Stone Age and take out billions of people in the process. This is not to suggest that humans should live in fear of a possible event that may never come, or if it did, might be millions of years in the future.

Problems of a solar coronal mass ejection, solar energetic particles or cosmic radiation are in fact much more likely to occur on a cosmic time scale and could certainly be destructive on a similar scale. The Carrington event, for instance, was about 150 years ago versus the K-T event, which occurred 65 million years ago. Nevertheless, there are clearly a number of logical steps that could and should be taken.

Step number one is to develop a much clearer and more precise way of communicating with the general public on possible threats that might come from potentially hazardous asteroids or other cosmic bodies. The Torino Scale that was adopted at the international Unispace Conference provides a clear representation of threat levels from 1 to 10 that the public can easily understand and appropriately respond to if and when a possible threat is identified. It would be hoped that most potential threats would be at a threat level of under three and thus put into proper perspective. What is needed is more public education to let people and school children know that there are possible future threats and to learn not only the Richter Scale or the relative sizes of hurricanes or tornadoes but also the Torino scale.

The second step is to complete with the right astronomical observations from the ground and from space-sensing satellites an inventory of NEOs. The WISE (Wide-field Infrared Space Explorer) has provided valuable information—as has ground-based observations—but much more needs to be done to identify with some certainty that all potential risks have been cataloged. Surely we are smarter than the dinosaurs and can allocate needed resources to completing the inventory in coming years and to carry out these tasks by allocating only a modest amount of our space research budgets for this purpose.

The third step is to investigate better strategies for diverting a potentially harmful NEO from its orbit so as to avoid collision with Earth and to make sure that such diversion does not create a future hazard.

8. What are the top things we should be doing to better identify the risks from NEOs?

Clearly the top priority is to identify NEOs that are larger than 1,000 meters in size. This task, which is the easiest to undertake and serves to identify the largest threats, is essentially complete. As far as we know any potential collisions from NEOs of this size are hundreds of years in the future. But there is legitimate concern about NEOs that are sized in the 100–1,000 m range. Here, more systematic observation needs to be completed since only about 20% of those objects in the 100–1,000 m range are estimated to have been detected.

The sensing capabilities of the WISE has accomplished a great deal in this respect, even though it was designed to detect much more distant cosmic objects through their infrared signature. Unfortunately the WISE satellite has now shut down since it can no longer function properly. In order to complete the inventory of these smaller NEOs one would need a WISE-type satellite with higher resolution and the ability to shift from wide range to narrow range focus on command and for this satellite to be essentially devoted to potentially harmful asteroid (PHA) detection.

Clearly the larger class of asteroid (i.e., those above 1,000 meters in diameter) would do much greater harm, but a 250 to 600 meter class PHA such as Apophis, if it were to hit Earth, could have the impact of tens of thousands of atomoic bombs. In short there are many more of the smaller class potentially hazardous asteroids out there, and an estimated 80 % of them are still undetected. Only one of these would still do enormous harm.

Finally, it would be useful not only to know the orbits of these satellites but to know their shape and composition in case it became necessary to seek to divert them from a collision with Earth.

9. What are the best strategies of coping with a potentially hazardous asteroid or other dangerous cosmic body that threatens Planet Earth.

There are studies and programs now under way to seek solutions to the problem of potentially hazardous asteroids. These activities can be generally sorted into the following areas: (i) Finding those potentially hazardous asteroids and comets that could most likely intersect with Earth orbit. This means not only those PHAs above the 1,000 m in diameter but also those in the range of 100–1,000 m that still remain 80 % unidentified. The NEO-WISE project has borne many results, but new space capabilities are needed to complete the task. (ii) Exploring the best strategies that could be employed to divert a PHA from Earth impact in terms of effectiveness and cost efficiency. Projects such as Earth Guard and NEOShield are often well conceived but are seriously underfunded. The identification of threats at the earliest possible time is key, because corrective action is most effective when carried out at as soon as possible; (iii) The final step is public education as well as that of legislators and government officials to help achieve an better understanding the various levels of threats from near-Earth objects.

10. Is there a systematic set of strategies that we could and should undertake with regard to manmade or cosmic threats in space?

The true key forward is an international space initiative that combines all of the resources of the world's space agencies to address space-related threats. There has already been good international progress made by the Inter-Agency Space Debris Coordination Committee (IADC), the U. N. Committee on the Peaceful Uses of Outer Space (COPUOS) and even the Space Data Association (SDA), but this effort is largely concentrated on the issue of coping with space debris and the launch of nuclear power sources into space. This effort needs to be expanded as part of the COPUOS new initiative on the "Long Term Sustainability of Space" [35]. There should be a clear global plan forward that addresses potentially hazardous NEOs, the cracks in Earth's geomagnetic shield, protection from the Sun's radiation, coronal mass ejections and solar energetic particles. "This would a global initiative to undertake a systematic planetary defense that can save the human species from extinction. Let's prove we are smarter than the dinosaurs."

About the Author

Dr. Joseph N. Pelton, Ph.D., is principal of Pelton Consulting International. He is the immediate past President of the International Space Safety Foundation, as well as chair of the Academic Committee and a member of the Executive Board of the International Association for the Advancement of Space Safety. He is the former Dean of the International Space University and Director Emeritus of the Space and Advanced Communications Research Institute (SACRI) at George Washington University. Dr. Pelton served as Director of the Accelerated Masters Program in Telecommunications and Computers at the George Washington University from 1998 to 2004. Dr. Pelton was the Director of the Interdisciplinary Telecommunications Program at the University of Colorado from 1988 to 1997, and at the time it was the largest such graduate program in the U.S. He previously held various positions at Intelsat and Comsat including serving as Director of Project SHARE and Director of Strategic Policy for Intelsat. Intelsat's Project SHARE gave birth to the Chinese National TV University.

Dr. Pelton was the founder of the Arthur C. Clarke Foundation and remains as the Vice Chairman on its Board of Directors. He has been active in the Arlington, Virginia community for many years as President of the Arlington County Civic Federation, as a member of the Long Range Planning Commission that initiated "smart growth" in Arlington and is currently Chairman of the IT Advisory Commission for Arlington County and Chair of the Civic Federation's Environmental Committee.

Pelton is widely published with some 35 books written, co-authored or co-edited. His *Global Talk* won the Eugene Emme Literature Award and was nominated for a Pulitzer Prize. He is the co-author of the books *Future Cities* (2009) and *The Safe City* (2013). These books examine how broadband communications can make cities safer and more responsive to the needs of citizens and improve education and health care services. Most of his books are about space, satellite communications, and the future of technology and its impact on society. He is on

J. N. Pelton, *Space Debris and Other Threats from Outer Space*,
SpringerBriefs in Space Development, DOI: 10.1007/978-1-4614-6714-4,

the Advisory Board of the World Future Society and also frequently speaks and writes as a futurist.

Dr. Pelton is a member of the International Academy of Astronautics, an Associate Fellow of the American Institute of Aeronautics and Astronautics (AIAA) and a Fellow of the International Association for the Advancement of Space Safety (IAASS). He was the Founding President of the Society of Satellite Professional International (SSPI) and a member of the SSPI Hall of Fame. In 2005 he won the ISCe Award for excellence in education and has also won the International Communication Association (ICA) award as the educator of the year. For the last two years he has served as President of the Comsat Alumni and Retirees Association .

He received his degrees as follows: B.S. from the University of Tulsa, M.S. from the New York University and his doctorate from Georgetown University.

Appendix

Acronyms and Key Terms

ACE	Advanced composition explorer spacecraft of NASA to obtain data from the sun's solar wind
AI	Artificial intelligence
AGI	Analytic Graphics Inc.
ASI	Agenzia Spaziale Italiana
Bolite	French for large-scale meteorite
CME	Coronal mass ejection
CNES	Centre National d'Etudes Spatiales, the French space agency
CNSA	China National Space Administration
COPUOS	Committee on the Peaceful Uses of Outer Space
CSA	Canadian Space Agency
DLR	German Aerospace Center
ESA	European Space Agency
EDDE	Electro-dynamic debris elimination
GEO	Geosynchronous earth orbit
GBL	Ground-based laser
IAASS	International Association for the Advancement of Space Safety
IADC	Inter Agency Space Debris Coordinating Committee
INREMSAT	Proposed international organization. The acronym would stand for: International Removal, Maintenance and Servicing of Satellites
ISRO	Indian Space Research Organization
ISSF	International Space Safety Foundation
JAXA	Japan Aerospace Exploration Agency
LEO	Low earth orbit
MEO	Medium earth orbit

J. N. Pelton, *Space Debris and Other Threats from Outer Space*,
SpringerBriefs in Space Development, DOI: 10.1007/978-1-4614-6714-4,
© Joseph N. Pelton 2013

NASA	National Aeronautics and Space Administration
NEA	Near earth asteroid
NEO	Near earth orbit
NOAA	National Oceanic and Atmospheric Administration of the United States. This agency works closely with NASA to monitor the solar wind and coronal mass ejections
NSAU	National Space Agency of Ukraine
ROSCOSMOS	Russian Federal Space Agency
PHA	Potentially hazardous asteroid
SBUV Radiometer	Solar backscatter ultra violet radiometer. These are space weather sensing devices on NOAA satellites
SDA	Satellite Data Association
SEP	Solar energetic particles that are associated with solar flares
SOHO	The solar and heliospheric observatory spacecraft, a joint undertaking between NASA and ESA to study the Sun and particularly its coronal mass ejections
SSS	Space surveillance system
UNO	United Nations Organization
USAF	U. S. Air Force
WISE	Wide-field Infra-red Space Explorer satellite

Bibliography

1. K-T event. Jet Propulsion Lab. http://www2.jpl.nasa.gov/sl9/back3.html.
2. Launch of Sputnik on Oct 4, 1987. www.history.nasa.gov/sputnik/.
3. Logsdon, John. 2010. *John F. Kennedy and the race to the moon.* New York: Palgrave-MacMillan.
4. Goddard Robert. The moon man. http://www.legacy.com/ns/news-story.aspx?t=robert-goddard–the-moon-man&id=279.
5. Space.com Staff Report. 2012. New debris-tracking 'space fence' passes key test. Space.com. http://www.space.com/14867-space-fence-orbital-debris.html. Accessed 12 March 2012.
6. Moskowitz, Clara. 2011. Space junk problem is more threatening than ever, Report Warns. Space News. http://www.space.com/12801-space-junk-threat-orbital-debris-report.html. Accessed 1 Sep 2011.
7. The Looming Space Junk Crisis: It's Time to Take Out the Trash *Wired Magazine.* www.wired.com/magazine/2010/05/ff_space_junk/all/1. Accessed 24 May 2010.
8. Op cit, Clara Moskowitz, "Space Junk…."
9. Liou, J-C., and N.L. Johnson. 2006. Risk in space from orbital debris. *Science* 311: 340–341.
10. NASA site on Near Earth Objects. http://neo.jpl.nasa.gov.
11. Solar Cycle Progression and Prediction. NOAA. http://www.swpc.noaa.gov/SolarCycle/
12. New Debris-Tracing 'Space-Fence' Passes Key Test. http://www.space.com/14867-space-fence-orbital-debris.html. Accessed 12 March 2012.
13. Space Fence Mark II: Prototype S-Band Radar Track Space Debris: http://www.gizmag.com/space-fence-radar-detects-debris/21779/.
14. Pelton, Joseph N. 2012. A fund for global debris removal. As presented at the International Association for the Advancement of Space Safety (IAASS) Conference in Versailles, France, Nov 2011 and International Space University Symposium on Space Debris, March 2012.
15. Tomasso Sgobba. IAASS Study on Space Debris Remediation: An Operational and Regulatory Framework for Assured Debris Removal. Nov 2011.
16. The Inter-Agency Space Debris Coordinating Committee. http://www.iadc-online.org/index.cgi.
17. IADC Space Debris Mitigation Guidelines. http://www.iadc-online.org/index.cgi?item=docs_pub.
18. Space Data Association. http://www.space-data.org/sda/about/members/.
19. Pearson, Jerome, Eugene, Levin, and Joseph Carroll. 2011. Commercial space debris removal. *Space Safety Magazine* (1): 21–22.
20. Proceedings of the International Interdisciplinary Congress on Space Debris, May 7–9, 2009 http://www.mcgill.ca/channels/events/item/?item id=104375 also see David Kushner.

J. N. Pelton, *Space Debris and Other Threats from Outer Space,*
SpringerBriefs in Space Development, DOI: 10.1007/978-1-4614-6714-4,
© Joseph N. Pelton 2013

2010. The future of space: orbital cleanup of cluttered space. *Popular science* Aug 2010, 60–64 and see: Joseph N. Pelton. 2012. The problem of space debris. *The Fundamentals of Satellite Communications*. 29–33. New York: Springer Press.

21. Russia to Spend $2 Billion For Space Clean Up. Space Daily. 2010. http://www.spacedaily. com/reports/Russia. Accessed 10 Nov 2010.

22. Bombardelli, Claudio et al. Dynamics of ion-beam propelled space debris. http://web. fmetsia.upm.es/ep2/docs/publicaciones/ahed11a.pdf.

23. A Super Solar Flare. NASA science news. http://science.nasa.gov/science-news/ science-at-nasa/2008/06may_carringtonflare/.

24. Hadhazy, Adam. 2009. A scary 13th: 20 years ago earth was blasted with a massive plume of solar plasma. Scientific America. http://www.scientificamerican.com/ article.cfm?id=geomagnetic-storm-march-13-1989-extreme-space-weather. Accessed 13 March 2009.

25. Fox, Nicky. 2011. Coronal mass ejection. Goddard Space Flight Center, NASA. http://www-istp.gsfc.nasa.gov/istp/nicky/cme-chase.html. Accessed 6 April 2011.

26. NASA-SOHO. www.nasa.gov/mission_pages/soho/.

27. NASA Stereo Satellite to Study Solar CMEs in Three Dimensions. http://www.nasa.gov/ mission_pages/stereo/.

28. The Solar Backscatter Ultra-Violet Sensor on NOAA Satellites (SBUV/2). http://www. ballaerospace.com/page.jsp?page=93.

29. The Earth's Magnetosphere Shield. http://science.nasa.gov/science-news/science-at-nasa/20 03/03dec_magneticcracks/.

30. Adams, Mike. 2011. Earth's magnetic pole shift unleashing poisonous space clouds lined to mysterious bird deaths. Natural News. http://www.naturalnews.com/030996_bird_deaths_ pole_shift.html. Accessed 13 Jan 2011.

31. Stewart, Iain, and John, Lynch. 2007. *Earth: The biography*, 57–63, Washington, D.C.: National Geographic Society.

32. Morrison, D., C.R. Chapman, D. Steel, and R.P. Binzel. 2004. Impacts and the public: Communicating the nature of the impact hazard. In *Mitigation of hazardous comets and asteroids*, ed. M.J.S. Belton, T.H. Morgan, N.H. Samarasinha and D.K. Yeomans. Cambridge: Cambridge University Press, 2004. This version reflects a revision of the Torino Scale. Also see. http://neo.jpl.nasa.gov/images/torino_scale.jpg for more details.

33. Pelton, Joseph N. 2012. Taking potentially hazardous asteroids (PHAs) seriously—making the public aware. Space Safety Magazine, Fall 2012.

34. Firth, Naill. 2010. Massive asteroid could hit earth in 2182. Warn Scientists. http://www.dailymail.co.uk/sciencetech/article-1298285/Massive-asteroid-hit-Earth-2182-warn-scientists.html. Accessed 28 July 2010.

35. U.N. Committee on the Peaceful Uses of Outer Space. Draft Report of the Working Group on the Long Term Sustainability of Space. http://www.oosa.unvienna.org/oosa/COPUOS/ ac105l.html.